PLANNING FOR RURAL RESILIENCE

PLANNING FOR RURAL RESILIENCE

COPING WITH CLIMATE CHANGE AND ENERGY FUTURES

EDITED BY WAYNE J. CALDWELL

UMP

University of Manitoba Press

IN MEMORY OF DR. RALPH KRUEGER. HE INSPIRED
A GENERATION OF PLANNERS AND GEOGRAPHERS AND
WAS A PERSONAL INSPIRATION, FRIEND AND MENTOR.

University of Manitoba Press
Winnipeg, Manitoba
Canada R3T 2M5
uofmpress.ca

Printed in Canada
Text printed on chlorine-free, 100% post-consumer recycled paper

19   18   17   16   15   1   2   3   4   5

Cover design: Frank Reimer
Interior design: Jess Koroscil

Library and Archives Canada Cataloguing in Publication

Planning for rural resilience : coping with climate change and energy
futures / Wayne J. Caldwell, editor.

Includes bibliographical references and index.
Issued in print and electronic formats.
ISBN 978-0-88755-780-4 (pbk.)
ISBN 978-0-88755-463-6 (PDF e-book)
ISBN 978-0-88755-461-2 (epub)

1. Regional planning—Ontario, Southern—Case studies. 2. Climatic
changes—Ontario, Southern—Case studies. 3. Sustainable agriculture—
Ontario, Southern—Case studies. 4. Ontario, Southern—Rural conditions—
Case studies. 5. Ontario, Southern—Environmental conditions—Case
studies. I. Caldwell, Wayne, 1957–, author, editor II. Reid, Susanna,
1967–. Building community resilience in Huron County.

HT395.C32O572 2015   307.1'212097132   C2014-907485-9
                                                     C2014-907486-7

The University of Manitoba Press gratefully acknowledges the financial support
for its publication program provided by the Government of Canada through the Canada
Book Fund, the Canada Council for the Arts, the Manitoba Department
of Culture, Heritage, Tourism, the Manitoba Arts Council,
and the Manitoba Book Publishing Tax Credit.

FSC
www.fsc.org
MIX
Paper from
responsible sources
FSC® C016245

TABLE OF CONTENTS

ILLUSTRATIONS

TABLES

ACKNOWLEDGMENTS

Many people contributed to this book. The contributing authors brought experience, insight, and passion to this topic. This book is also the product of a research project at the School of Environmental Design and Rural Development, University of Guelph. Numerous individuals contributed to this project and have influenced the book in one way or another. They include: Jennifer Ball, Erica Ferguson, Paul Kraehling, Émanuèle Lapierre-Fortin, Amy Lejcar, Eric Marr, Shannon McIntyre, Adam Wright, Colin Dring, and Jennifer Sisson. Funding for this project was provided by the Ontario Ministry of Agriculture, Food and Rural Affairs. Monika Rau at the University of Guelph helped with many details as this book progressed and staff at the University of Manitoba Press, including Jill McConkey and Glenn Bergen, and copyeditor Dallas Harrison, were most helpful.

INTRODUCTION: SETTING THE STAGE

WAYNE CALDWELL, ERICA FERGUSON, ÉMANUÈLE LAPIERRE-FORTIN, AND JENNIFER BALL

A "game changer" is defined as an event, idea, or procedure that effects a significant shift in the current way of doing or thinking about something. (*Oxford English Dictionary*)

One hundred years from now, it is highly probable that we will have faced two daunting, debilitating, and unpredictable challenges. Climate change and peak oil threaten the future of rural communities as we currently know them. Climate change is expected to play havoc with our weather, with corresponding impacts on the environment, the economy, and society. Although there is debate over the timing and impacts from peak oil (made all the more difficult because of the resurgence of production based on fracking), there will be a time when the availability and pricing of conventional oil will significantly impact life in rural communities. These two issues pose much uncertainty. As we look to the future, we can imagine a range of possible scenarios, from the positive to the negative and from the predictable to the uncertain. Although the future is neither certain nor guaranteed, this book embraces a hopeful future—one in which we have successfully responded to the challenges posed by climate change and peak oil.

This introduction establishes the context of and framework for this book. It reflects on relevant literature, introduces the concept of resilience, and discusses its relevance to the future of rural communities. It then discusses the impacts of climate change and peak oil and the joint and related implications for rural communities. As well, it highlights the application of resilience as a response to these impacts. Ultimately, the introduction establishes the foundation for the case studies, planning actions, and community-based strategies presented in the chapters.

INTRODUCTION

Almost all climate change and peak oil literature asserts that climate change is happening and suggests that peak oil is looming. Planning, as a future-oriented profession working for the public interest, must respond: "The implementation of adaptation measures is essential for municipal governments to protect the well-being of citizens and to manage public resources effectively" (Richardson 2010, 32). Diamond (2006) describes how some settlements in Greenland, Iceland, and Easter Island collapsed because of the mismanagement of natural resources and the inability to adapt to environmental changes. However, societies can create resilient communities that use less oil, which, according to Newman, Beatley, and Boyer (2009, 11), could increase quality of life, reduce impacts on the environment and human health, result in greater equity and economic gain, lessen economic vulnerability, and contribute to a more peaceful geopolitical context. In a world characterized by rapidly occurring changes, peak oil and climate change are two drivers that have far-reaching implications for societies reliant on fossil fuels. This is the context in which "community members [can] collectively and strategically engage their resources to respond to change" (Magis 2010, 405) by increasing community resilience.

RESILIENCE TO CLIMATE CHANGE AND PEAK OIL

Climate change adaptation specialists Nelson, Adger, and Brown (2007) argue that resilience provides a useful framework for analyzing climate change adaptation processes and identifying appropriate policy responses. Resilience is a multifaceted and dynamic concept; as Coutu notes (2002, 42), "resilience is one of the greatest puzzles of human nature." In fact, there is a plurality of definitions of resilience from different disciplinary fields (Gallopin 2006, 293), ranging from mental health to public health, disaster response, socio-ecological systems, community development, and natural resource management.

The concept of resilience originates in ecology, where it generally refers to the ability to withstand stress and recover quickly. Nelson, Adger, and Brown (2007) describe the resilience of socio-ecological systems as the amount of change that a system can undergo while retaining the same function, structure, and options to develop. In the planning profession, building resilience is most often conceptualized as part of disaster preparedness, and resilience often responds to specific or one-off types of natural disaster, such as floods, tsunamis, earthquakes, and so on. For example, Godschalk (2003, in Ahmed 2006) suggests characteristics of resilient systems that can create disaster-resilient cities:

> *redundancy*—systems designed with multiple nodes to ensure that failure of one component does not cause the entire system to fail;
>
> *diversity*—multiple components or nodes versus a central node, to protect against a site-specific threat;
>
> *efficiency*—positive ratio of energy supplied to energy delivered by a dynamic system;

autonomy—capability to operate independent of outside control;

strength—power to resist a hazard force or attack;

interdependence—integrated system components to support each other;

adaptability—capacity to learn from experience and flexibility to change; and

collaboration—multiple opportunities and incentives for broad stakeholder participation.

The literature suggests that the concept of resilience building can be broadened to include adaptation to other types of phenomena that are more long term, such as those brought about by climate change and peak oil. According to Newman, Beatley, and Boyer (2009), planners can use technologies such as small-scale water, waste, and renewable energy systems, biomimicry, green chemistry, and industrial ecology to rethink communities and move toward more localized, polycentric, distributed, and eco-efficient approaches that prepare for climate change and the end of cheap oil. Daniel Lerch (2009), the program director at the Post Carbon Institute, argues that, "to truly build the resilience of all our communities against the coming changes in the global oil supply, urban planners and policymakers will need to turn aggressively to more systems-informed approaches to community governance and development."

COMMUNITY RESILIENCE
Consensus is emerging that building resilience at three nested levels—psychological/personal, community, and system—must be at the centre of convergent social justice and environmental social change movements (Poland et al. 2009). Others (Norris et al. 2008) describe community resilience as a process of linking a network of adaptive capacities to adaptation after a disturbance or adversity. These capacities include economic development, social capital, information and communication, and community competence.

Change is a key word in the literature on community resilience. Magis (2010) proposes that change is the main factor that distinguishes community resilience from the more well-known concept of community capacity. Resilience is increasingly important in planning, for change is a constant and inevitable feature of modern societies with the uncertainties brought about by climate change and peak oil. Resilient communities can "bounce back" in the face of change. Change, however, is not inherently negative or positive, and planners can work with communities to effect this change: "Resilient communities do not simply withstand or react to external pressures; rather, they initiate positive change and are enriched and unified in the process. They embody the idea that no matter where you live, or whatever constraints you face in terms of location or resources, you can, with the right support, create a vibrant, sustainable and equitable community, which should defy even the most destructive impacts of social, economic, political, technological and economic change" (Carnegie Commission for Rural Community Development 2007, in MacDonald 2009, 1).

In addition to responding to change, resilience reduces vulnerability and builds adaptive capacity (Gallopin 2006; Magis 2010). Many authors perceive resilience as a state either achieved or not. However, others argue that community resilience is a process, not a goal. In constructing a model based on mixed methods research in three resource-dependent rural communities in western Canada, Kulig, Edge, and Joyce (2008, 77) conceptualize community resilience as a process. That process can be seen as a flow chart in which a "sense of belonging" begets the expression of a "sense of community" and togetherness, which in turn begets "some type of community action, noted by the presence of visionary leadership, an ability to deal with change in a positive way, an ability to cope with divisions, and the emergence of a community problem-solving process." This research aligns with the findings of Maybery et al. (2009) that social and community connectedness are crucial determinants of community resilience and well-being—that the social aspects of community resilience are crucial.

Actions can be taken to increase resilience in any community. In developing community resilience, some critical elements compiled by Callaghan and Colton (2008, 932) include planning and developing strategies that minimize vulnerabilities, developing communication and crisis response systems, supporting government/private partnerships and independent initiatives that create social support, and developing strategies that diversify risk across space, time, and institution. Also important is nurturing diversity to enable post-shock reorganization and renewal and to enable parties with different kinds of knowledge to self-organize and engage in social learning (Berkes and Seixas 2005, in Magis 2010). Resilient communities have thus been described, by a group seeking to increase resilience in the San Francisco Bay area, as communities that use their assets in creative ways to withstand and recover from hard times, meet basic human needs, and show strength and creativity, no matter what the circumstances (Bay Localize 2009). To "bounce back," resilient communities are proactive, not reactive. In fact, Colussi et al. (1999, 11), in a Canadian Centre for Community Renewal publication intended to guide communities toward increased resilience, define a resilient community as "one that takes intentional action to enhance the personal and collective capacity of its citizens and institutions to respond to ... and influence the course of change."

Magis (2010, 401), after conducting an extensive literature review on community resilience, defines it as follows: "the existence, development, and engagement of community resources by community members to thrive in an environment characterized by change, uncertainty, unpredictability, and surprise. Resilient communities intentionally develop personal and collective capacity to respond to and influence change, to sustain and renew the community and to develop new trajectories for the community's future."

Magis (2007) develops an Index of Community Resilience that presents five key constituents. They are discussed later in this book in the case study of Eden Mills Going Carbon Neutral (EMGCN). Magis argues that social and physical infrastructure must be in place to provide community space in which to gather, learn, and collaborate. In addition, the community needs financial resources, collective knowledge, skills, and abilities to anticipate and respond to change, and a diversity of community members actively engaged in strategic community planning. This can be boiled down to five key areas of community resources or capital (social, physical/financial, human, political, and cultural) that need to be present, developed, and engaged in resilient communities.

COMMUNITY BUILDING IN RESILIENT COMMUNITIES

Numerous authors have mentioned different aspects of community building as a pinnacle element of community resilience. Breton (2001, in Kulig et al. 2008, 77) has noted that resilience is dependent on neighbour networks and active local voluntary associations. Szerzvniski (1997, in Balls 2010, 18) argues that community-led movements are now as important as governmental top-down action in terms of practical sustainability. Narrowing in on the social side, Adger (2003) defines social resilience as the ability of groups or communities to cope with external stresses and disturbances as a result of social, political, and environmental change. He suggests that different sources of social resilience include social networks, lessons learned from past experiences, high diversity, learning through consensus building, and strong support networks. He also argues that "building trust and cooperation between actors in the state and civil society over [climate change] adaptation has double benefits. First, from an instrumentalist perspective, synergistic social capital and inclusive decision-making institutions promote the sustainability and legitimacy of any adaptation strategy. Second, adaptation processes that are built from the bottom up and are based on social capital can alter the perceptions of climate change from a global to a local problem" (ibid., 401). Colussi (2000) has observed that social and economic development organizations in resilient communities work to inform and engage the public and demonstrate high levels of collaboration with each other.

To be resilient themselves, Anderson (1994, in Cadell, Karabanow, and Sanchez 2001, 27) states, organizations need to possess five characteristics: (1) a clear mission, (2) shared decision making, (3) trust building, (4) the encouragement of openness, and (5) the enhanced competence of the individual and the collective. Social aspects of community resilience relevant to the work of community organizations include engagement, education, empowerment, and encouragement (Edwards 2008, 11).

Callaghan and Colton (2008, 940) argue that ultimately the success of building a sustainable and resilient community depends on strong leadership, vision, and clear and open communication. Here again the multifaceted character of issues at hand, and the diversity of players involved, call for innovative responses, namely a new kind of leadership (leading between?) focused on building collaborative relationships built on mutual respect rather than formal authority; furthermore, social innovation will be an important aspect needed in this context (Torjman 2007). Lerch (2007) advises local governments to do anything they can to get people talking with each other, forming relationships, and investing in the larger community.

In summary, if the arguments on the role of community building in creating resilient communities hold true, then planners and local governments need to respond by working with civil society. Well-connected citizens' organizations engaged in formal and informal partnerships with local governments committed to supporting grassroots community action can be an important piece of planning for resilient communities. However, those alliances are in early stages across the country, and much still remains to be done to advance potential alliances. For complex issues, it has been argued, local governments cannot be content to follow public opinion but must show leadership, engage partners, set the agenda, take risks, and be role models (Noble and Abram 2008, 13). The question remains whether or not local governments have the capacity to achieve such a level of leadership on climate change and peak oil with their current level of resources and expertise. The following section suggests how municipalities can show such leadership.

PLANNING FOR RESILIENT COMMUNITIES

Planners and local governments are uniquely positioned to respond to the challenges and implications of climate change and peak oil: though "resilience theory lends itself to many disciplines, few are better suited to the approach than land use planning" (Environmental Commissioner of Ontario 2009). The potential implications include strained global water resources, volatile food supplies, economic contraction, increased conflict over key natural resources, and strained ecosystems (Bay Localize 2009, 22). Local governments have influence over three key areas of spatial and economic development: building construction and energy efficiency (through zoning by-laws, building codes, and permit processes), local land-use and transportation patterns (mobility choices), and local economic activity that creates opportunities to encourage development in low-energy, zero carbon directions by both incentive and example (Lerch 2007, vi). It is in every municipality's best economic interest to act on these issues, because higher-level governments cannot see the details that local governments can and are often slow to respond (ibid., 27). While no easy task, building resilience to climate change and peak oil is not impossible: "Although the climate crisis is new, many of the underlying challenges are not. Local governments have dealt with other environmental and social problems in the past. They have lots of experience managing the 'messiness' and getting it done. They know the importance of leadership, of successful change management, of innovation and risk-taking, of getting started. Much of this experience is relevant to the climate change challenge" (Noble and Abram 2008, 13).

There are encouraging and effective ways that planners are responding to the impacts of climate change and peak oil; one can think of Smart Growth policies and the Healthy Communities movement, as well as the subfield of disaster preparedness, as examples of planners confronting interrelated, complex issues. Research by Harvey (2009) and Lerch (2007) has shown that mitigation measures with the greatest potential are related to increasing energy efficiency, reducing energy consumption, and producing fossil fuel–free renewable energy locally. All of these are areas that planners can influence through guidelines for land-use planning that make energy use and production the principal determinants in land-use decisions, resulting in favouring "brownfield" over "greenfield" development, planning for mixed use and intensification, fostering vibrant centres, and ensuring low-energy transport (Gilbert 2006, 38). In addition, according to the UK Sustainable Development Commission (Owen 2009), social-planning policy supporting a diverse local service base, as well as land-use planning policy supporting the creation of public space for community involvement and neighbourliness, are two important features of a resilient community.

Any element of building resilience to climate change and peak oil must include creating and implementing a strategic plan. Lerch (2007) advises councils aiming for resilience to commit publicly to action by endorsing the World Mayors and Municipal Leaders Declaration on Climate Change, joining ICLEI's—Local Governments for Sustainability Cities for Climate Protection Campaign, signing the Oil Depletion Protocol (which sets a target for reducing oil consumption across the community), and establishing a Peak Oil Task Force to identify the challenges and vulnerabilities that the community faces as a result of peak oil and climate change. These steps should also increase public awareness of likely impacts of peak oil and climate change and engage stakeholders in identifying problems and solutions.

PLANNING FOR RESILIENT COMMUNITIES: A CANADIAN PERSPECTIVE

In 2009, Thomas Homer-Dixon and Nick Garrison published *Carbon Shift: How the Twin Crises of Oil Depletion and Climate Change Will Define the Future*, a collection of Canadian articles on what many consider to be the most challenging topics of our time: climate change and peak oil. It is evident in this collection of essays that those at the forefront of combined knowledge of these twin crises are anything but united. Despite the controversy over which "crisis" is most pressing, the contributing authors concur that change is necessary on both fronts, that communities must acknowledge the inevitability of change, and that action is necessary. *Carbon Shift* highlights the broad difference of opinion and analysis in academic and industrial circles in Canada. For example, it includes Mike Jaccard's[1] essay "Peak Oil and Market Feedbacks," which argues that there is enough oil to last at least 100 years. It also includes William Marsden's[2] essay "The Perfect Moment," which offers this interpretation of the data: "Our [global] reserves total about 1.2 trillion barrels, give or take a few hundred million. We use this oil up at a rate of more than thirty billion barrels each year. That means that world reserves will last about another thirty-nine years. Petroleum geologists, such as Andrew Miall at the University of Toronto, predict that what's left undiscovered will give us another ten to twenty years, tops" (2009, 155). Furthermore, David Keith[3] contributed the essay "Dangerous Abundance," which asserts that, "while it is possible to make a case that oil scarcity poses a threat to our civilization, I argue that fossil-energy abundance is where the more urgent threat lies" (2009, 28). In other words, because of human-induced climate change, continuing to emit greenhouse gas by burning fossil fuels is more dangerous than the threat of running out of them.

In *Carbon Shift*, Canadian experts agree on two facts: that oil will eventually run out and that burning fossil fuels is causing climate change, which is having grave effects on the entire globe. In general, those arguing that there is plenty of oil remaining tend to be economists, and those most concerned about supply are geologists. This divergence of expert opinion is noted by Hopkins (2006), who suggests that economists tend to believe that market forces will manage the potential oil depletion crisis in that, as reserves are depleted, the price of oil will rise, which in turn will support further innovation and technology to increase production. Geologists, however, seem to have a view different from that of the marketplace that highlights the ultimate limitation of oil availability (ibid., 15). Regardless, peak oil is a theory that is currently contentious, and, if correct, its implications threaten all aspects of the current Canadian lifestyle.

AREA OF IMPACT	EFFECTS
TEMPERATURE AND WEATHER	Temperature increases; since 1948, temperatures in Ontario, for example, have increased up to 1.3 degrees Celsius in the western portion of the province. Acceleration of warming is expected, especially in the northern portion. More moisture and a warming atmosphere increase the likelihood of extreme weather events as well as more variable and less predictable weather.
ECOSYSTEMS	Habitat and biodiversity will be altered. There will be declines in forest density and area in some regions. Arid lands will expand in size. Implications for forestry include the spread of disease and insects as well as different growing conditions for trees.
WATER RESOURCES	Water shortages have occurred in southern regions and might become more frequent with higher temperatures and more evaporation. Changes to water quality and quantity will occur. There will be increased winter and spring runoff and decreased summer soil moisture. Severe floods and droughts will be more common.
FOOD AND FIBRE	Crop and livestock production will be influenced by changing temperatures, length of the growing season, rainfall, extreme weather, snow cover, and frost. There will be direct influences on pests, invasive species, weeds, and disease, as well as acid rain and smog. There will be greater susceptibility to changes because of water quality, pests, diseases, and fires. Some regions' crops will benefit from increased temperatures. The availability of water for irrigation might be an issue with changing rain patterns. There might be beneficial implications for crop production, potentially expanding production north.
COASTAL SYSTEMS	There will be a shift from cold/cool water fish species to warm water species. The rise in sea level will expand floodplains and destroy coastal wetlands. Decreases in the Great Lakes might reduce loads carried by freighters, and there might be impacts on hydroelectricity generation.
INFRASTRUCTURE AND HUMAN SETTLEMENT	Infrastructure faces risks from severe storm impacts (floods, road washouts, ice damage, windstorm damage) to softening of tarmac in summer and cracking of concrete in freezing and thawing. Wastewater and storm water infrastructure might not be adequate. Northern communities are vulnerable to earlier spring melts.
HEALTH	There will be health implications from heat waves, smog, and mosquito/tick-borne diseases, increasing the risk of illness and premature death. There might be adverse effects on human health as a result of direct causes, such as temperature stress, air pollution, and extreme weather. Potential adverse effects on human health as a result of indirect causes include vector-, rodent-, and water-borne diseases and exposure to toxic substances.

Table 0.1. Potential climate change impacts

CLIMATE CHANGE IN RURAL CANADA

The effects of human-induced climate change should by no means be diminished by the examination of peak oil: the two are inextricably linked, and both warrant the attention of policy makers, the media, and the public. The anticipated impacts as suggested in the literature are summarized in Table 0.1.

Human-induced climate change is now commonly perceived as a fact: despite opposition, it has moved beyond the realm of theory and been accepted as real. Current Canadian research identifies the status of climate change as "the paramount global environmental issue," the discussion of which "has moved from acceptance and attenuation ... to seeking an understanding of the implications of change and potential adaptation and resiliency strategies" (Community Research Connections 2010).

In 2003, the Canadian Standing Senate Committee on Agriculture and Forestry released a report entitled *Climate Change: We Are at Risk*. The summary of Chapter 5 identifies key considerations of climate change for the rural Canadian context:

> Because rural Canada relies largely on natural resource-based industries, it will be more vulnerable to climate change. Over the past several decades, rural communities in Canada have been changing dramatically, due to migration and structural transformations in resource-based industries. The livelihoods of rural Canadians are already stressed by low commodity prices and by trade conflicts such as the softwood lumber dispute and climate change will bring additional challenges, which may aggravate the current situation. Climate change will have significant financial and economic repercussions on natural resource-based industries, and physical infrastructure will also be challenged by increased weather-related damage. In order to cope with these changes, rural communities will have to start considering climate change effects in their planning.

In this vein, the Canadian Climate Impacts and Adaptation Research Network, in existence from 2001 to 2007, was specifically tasked to research climate change impacts and adaptation in Canada. Its online agriculture archive (http://www.c-ciarn.uoguelph.ca/documents/index.html) contains fourteen research papers pertinent to the risks of climate change for rural Canada. These types of initiatives are vital to assisting rural Canada in mitigating and adapting to climate change. Similar research and attention have not yet begun to provide knowledge and strategic actions required for the possibility of energy scarcity.

PEAK OIL IN RURAL CANADA

A decline in energy availability will have repercussions for almost every community on earth. In assessing strategies for mitigation of and adaptation to a new reality, some will be applicable universally, and others will require a regional or local flavour, specific to urban, suburban, or rural challenges. There is relatively little literature detailing the rural implications of Peak Oil, and, though the identification of impacts is highly speculative, the Groundswell Group (2007) identified what it perceived to be key impacts. Highlights of this report have been summarized by Klug (2009, 9) in Table 0.2.

To address these key challenges, it is important to understand the various contexts in rural communities. Research on rural community strategies for resilience to peak oil is notably absent. Rural areas are of increased significance when assessing the repercussions for food supply. In the United States, it is estimated that producing one joule of food energy for consumption by humans currently requires approximately ten joules of fossil fuel energy (Grubb 2010). The situation in Canada is equally dependent on inputs (fertilizers, machinery, transport) for food production and consumption. Some analysts believe that the easiest first step is to retool management of the current system, which could increase efficiency to "double the amount of energy service we get from each barrel of oil" (Ayres and Ayres 2010, 2). Although this might partially alleviate the troubles caused by decline, local strategies for increasing resilience are essential.

Rural Canada, and rural areas in general, are particularly vulnerable to the effects of climate change (Reid et al. 2007; Wall, Smit, and Wandel 2007) and to fuel scarcity for essential elements as diverse as transportation and agricultural reliance on liquid fuels. The risks of climate change are becoming better understood, and both mitigation and adaptation strategies are being researched. Strikingly, the potential implications of energy scarcity have not yet hit the public or policy-making consciousness in Canada, despite some efforts in other nations (notably the United States, Australia, and the United Kingdom). Thus, the inevitable decline in liquid fuel production—commonly referred to as peak oil—warrants attention at all Canadian levels: grassroots, municipal, provincial, and national. White, in *Issues in Canada: Climate Change in Canada* (2010, 16), details current Canadian energy use as derived from the following sources, in these percentages: fossil fuels 70 percent, hydroelectric 11 percent, nuclear 11 percent, biomass (wood) 6 percent, other renewables 1.5 percent. White also postulates about the agricultural implications for small farms in Canada—an analysis that relates to the implications of both climate change and peak oil: "A large agricultural corporation may be able to spread its risk by diversifying out of the most threatened regions. However, this is not an option for small family

AREA OF IMPACT	EFFECTS
ECONOMY AND SOCIETY	Cost for everything will rise (electricity, fuel and heating, food, and other products) Food spending as percentage of income will rise Spending priorities will change; less money for luxury items; income spread thin between food and heating (likely to affect the poorest first, with increased deaths of elderly people during winter months) Ageing population in rural regions will require greater assistance, especially for those living in the most isolated areas
EMPLOYMENT	Lack of consumer income will affect big-box retailers as consumers no longer wish to drive long distances This will cause consumers, industry, agriculture, and commerce businesses to remake the retail business model People living in rural communities will have to limit their financial capacity to commute; retail employees, who are often amongst the lower paid within the community, will be hit hard, as well as those reliant on their vehicles for work, such as postal workers, care in the community, delivery personnel, etc.
ENERGY PRODUCTION	Distant houses may be abandoned; more people may start living in one house to save on energy costs Frequent blackouts may occur, causing stress on the local and regional economies; it may also have major psychological effects
FUEL AND TRANSPORT	Increased pressure on farmers to grow biofuels on their land could prevent them from becoming self-sufficient and leave them more vulnerable to inflation rates Rural areas less able to afford commuting, travelling, etc.
ENVIRONMENT	Less scope or money available for climate change mitigation, plus turning to coal as an energy source exacerbates climate change problem Woodlots under threat for fuel to heat homes
TOURISM	Less international tourism, perhaps a greater reliance on domestic tourism (shorter distances) Huge effect on economies of places that rely heavily on tourism
HEALTH SERVICES	Surgeries and hospitals will become less able to prescribe drugs and medicines to those in need Health systems and pharmaceuticals derived from petroleum will be in short supply (i.e., analgesics, antihistamines, antibiotics, cough syrups, creams, ointments, salves, radiological dyes and films, intravenous tubing, syringes, oxygen masks, heart valves, etc.)
POPULATION AND MIGRATION	Food shortages could put population pressure on rural areas as more people move to be closer to the food supply

Table 0.2. Potential peak oil impacts. Sources: Groundswell Report (2007); Klug (2009)

businesses, except to the extent that many farming families rely on off-farm employment to supplement their income" (ibid., 35).

It is becoming apparent in the current literature on the nexus of climate change and peak oil that the concept of peak oil not only provides support for initiatives to tackle climate change but also provides a pathway to examine and respond to the resulting challenges facing communities in addressing these "twin crises." Agricultural and rural communities, with a plethora of other challenges, are not alone in being ill equipped to face these challenges.

CLIMATE CHANGE AND PEAK OIL: DOUBLY CHALLENGING

Peñuelas and Carnicer's "Climate Change and Peak Oil: The Urgent Need for a Transition to a Non-Carbon-Emitting Society" (2010) clearly indicates the interplay of climate change and peak oil and the absolute necessity of addressing it. Climate change and peak oil are deemed to be "twin" issues because of their level of interconnectedness, because "oil use is responsible for approximately one-third of greenhouse gases" (Newman, Beatley, and Boyer 2009, 7). Both problems are complex and associated with a great deal of uncertainty. For example, it is possible that the costs of adapting to climate change might coincide with an economy strained from rising energy prices. One typical way in which planners handle such uncertainty is through scenario-building exercises, an emerging and dynamic field for climate change and peak oil. There is a wide range of scenarios in the literature, each with specific sets of implications.

Holmgren (2009) identifies four broad categories of scenarios: *techno-explosion*, which could go so far as to include colonizing other planets for resources; *techno-stability*, a favourite of those who argue that, once sustainable systems are in place, a steady-state society can flourish; *energy descent*, which involves a reduction in economic activity, complexity, and population as fossil fuels are depleted and a ruralization of settlement and economy begins; and *collapse*, a scenario characterized by a rapid decline in population and biodiversity loss. Holmgren argues that the mounting evidence on future climatic conditions and trends for global oil supply and demand suggests that the next energy transition will have transformative effects on the social and physical fabric of industrial society. His argument is based on the concept of net energy and energy quality; in short, no other source of energy has the same "bang for the buck" as fossil fuels, which makes continued and increasing supply for energy highly unlikely. Holmgren predicts that "some sort of energy descent [is] increasingly likely despite the deep structural and psychological denial of the evidence" (ibid., 28). Some authors, such as Curtis (2009), suggest that localization is a peak oil mitigation and adaptation strategy. He uses the term "peak globalization" to explain how peak oil and climate change will undermine globalization by creating conditions for increased prices of goods and transportation, which will likely result in shorter supply chains that encourage goods to be produced close to where they are consumed.

Friedrichs published a compelling piece in *Energy Policy* in 2010 entitled "Global Energy Crunch: How Different Parts of the World Would React to a Peak Oil Scenario." He develops in this article a historical basis for analyzing our current situation through examples of national peak oil scenarios in the twentieth century by identifying three paths that nations have taken during times of fuel scarcity: predatory militarism, based on the Japanese experience from 1918 to 1945; totalitarian retrenchment, as exhibited by North Korea in the 1990s; and socio-economic adaptation, using Cuba's 1990s "Special Period" as an example. His analysis and hypotheses about the future arise from these case studies. Friedrichs is hesitant to assert one point of view on peak oil theory, but he believes that forecasting how different parts of the world might react is a timely and necessary venture.

Despite wide recognition of the peak oil concept and its linkage to climate change, the nature of their relationship continues to be the subject of international debate. Some argue that the two challenges are similar in their cumulative effects on the economy, where the occurrence of peak oil and rising fuel prices will be delayed by increased carbon taxes aimed at mitigating climate change or where they will both affect the comparative advantage currently experienced by the global supply chain (Curtis 2009; Partridge 2007). Others argue that they are linked through the effects of fossil fuel consumption on climate change, much of which is the result of global transportation and energy-intensive industries; it is further argued that they are linked in the intersection of the end of "cheap" oil and the effects of climate change (Heinberg 2007; Kunstler 2006; Mehdi, Mrena, and Douglas 2006). The commonalities of these linkages are that the timeline for effects on society is unknown and impending, that institutions and organizations will have to consider these issues in tandem when making decisions for the future, and that much academic analysis of climate change has omitted an analysis of peak oil (Curtis 2009; Heinberg 2007; Reynolds 2010).

The implications of all this for rural communities are troubling. Although rural communities have the advantage of space and access to certain resources, they also tend to lack the human and capital resources that serve large concentrations of urban populations. We can be relatively certain that primary industries such as agriculture, forestry, and fisheries will exist in a time of uncertainty. Although some forecasts suggest enhanced agricultural production in parts of rural Canada, there is also recognition that the growth in agricultural productivity in the past has been based on inexpensive supplies of energy. As we link climate change and peak oil, there is a certainty of uncertainty. In this context, planning for resilience becomes a desirable if not essential option.

PLANNING FOR RESILIENCE

Canadian communities are still generally in an early stage of building resilience in the face of climate change and peak oil. Natural Resources Canada (NRC, 2008), in a summary of key findings on climate change impacts and adaptation in the country, indicates that climate-related disruptions to critical infrastructure, water shortages, and climate-related extreme events have occurred in Canada and are likely to become increasingly frequent in the future. The impacts of these present risks will affect the health of Canadian residents and ecosystems. In Canada, projects responding to such impacts might be eligible to receive funding from the federal government's $550 million Green Municipal Fund program, managed by the Federation of Canadian Municipalities (FCM), which, along with FCM's Centre for Sustainable Community Development, are two national efforts to support sustainable development at the municipal level. Municipalities in most provinces across Canada are obliged to prepare "Integrated Community Sustainability Plans" as part of the Federal Gas Tax Agreement, under which municipalities must demonstrate progress toward enhanced sustainability planning by 2010 in exchange for funds received (AMO 2008).

At the municipal level, there are many ways that councils and planners can contribute to a low-carbon future (Rowell 2010). *Municipal World* has compiled case studies of Canadian municipal responses to climate change (Gardner and Noble 2008); cities such as Burnaby, British Columbia, and Hamilton, Ontario, are conducting research and creating strategic plans in response to the potential impacts of peak oil and suggesting responses to reduce energy use in transportation and produce energy locally. In addition, the Canadian Institute of Planners (CIP, 2010) has published a national policy framework for climate change adaptation as part of a two-year project to mainstream climate change tools for the planning profession. The CIP policy on climate change seeks to increase planners' capacity to mitigate and adapt by facilitating networking as well as information and knowledge transfer on the links between planning and climate change.

Natural Resources Canada believes that certain provinces have a strong adaptive capacity[4] but that it is unevenly spread out and particularly low in rural communities. The adaptive capacity of rural communities hinges on strong social capital, social networks, attachment to community, strong traditional and local knowledge, and high rates of volunteerism, yet it is limited by economic resources, less diversified economies, higher reliance on natural resource sectors, isolation, limited access to services, and lower proportion of the population with technical training (NRC 2008, 13).

STRUCTURE OF THE BOOK

This introduction provides a context for this book. It points to the concept of resilience as an appropriate framework for responding to the twin challenges of peak oil and climate change. Even though the impact and timing of these two challenges remain somewhat speculative, there is no doubt that rural Canada needs to consider, prepare for, and respond to the impacts. The focus on rural communities points to the specific challenges that such communities face. These challenges reflect the combined effects of geography, the environment, the economy, and the resulting livelihoods of rural residents.

This book includes ten chapters with a range of issues from natural disasters to agriculture. The first chapter, by Susanna Reid, establishes a theoretical framework to help understand the impacts of a natural catastrophe (a tornado) and the issue of resilience both in relation to the community and in relation to economic activities. The chapter speaks to the roles of individuals, households, and businesses as well as planning processes (land-use, community, but also strategic planning). The second and third chapters, by Paul Kraehling and Wayne Caldwell, look at the role of nature and resilience, linking development with nature and conservation. Green infrastructure is profiled as an asset that can help communities on the path to resilience. The fourth chapter, by Eric Marr, investigates rural transportation. The chapter speaks to the dependency of rural communities on transportation and in turn identifies the vulnerability associated with reliance on the automobile and fuel prices. Public transportation is identified as an option, and examples from elsewhere in the world are presented. Chapters 5 and 6 profile the experiences of two communities: Eden Mills, which has adopted a carbon-neutral strategy, and Guelph, which has pursued the Transition Town movement. These chapters consider the impacts of rising oil prices and the relationship to community resilience. They consider the roles of different actors, social capital, and community solidarity. Each chapter concludes with an interesting SWOT (strengths, weakness, opportunities, threats) analysis and demonstrates that one does not need a hierarchical structure to be effective. Chapters 7, 8, and 9 focus on the agricultural sector. Chapter 7, by Erica Ferguson, presents an interesting set of case studies profiling the role of organizations in supporting food production in the context of community resilience. Chapter 8 takes a somewhat different approach and presents the reflections of an individual farmer. Tony McQuail profiles his career in farming, explaining why he rejected elements of the current agricultural system and eventually adopted an organic approach to farming integrated within a strong community context. Chapter 9 identifies a number of challenges for agriculture related to food security. The chapter discusses productivist agriculture and its negative consequences and lack of adaptation to climate change. The literature on adaptation identifies various types and processes of adaptation. Finally, Chapter 10, by Christopher Bryant, provides a summary focused on the concept of sustainability for rural communities. This chapter draws together many of the themes addressed by the other authors of this book. Presented is a view of community resilience based on local action and appreciation of the importance of community resilience while recognizing the essential heterogeneity of rural communities.

Notes

1   Jaccard is a professor in the School of Resource and Environmental Management at Simon Fraser University, Burnaby, British Columbia. He is the author of *Sustainable Fossil Fuels* (2005), which won the Donner Prize for best book on Canadian public policy.

2   Marsden is an author and documentary filmmaker. He has won numerous awards, and his most recent book *Stupid to the Last Drop*, which won the 2008 National Business Book Award, is an in-depth critical appraisal of the tar sands in Alberta.

3   Keith is a professor in the Department of Chemical and Petroleum Engineering and the Department of Economics at the University of Calgary, and he is an adjunct professor in the Department of Engineering and Public Policy at Carnegie Mellon. He has held the Canada Research Chair in Energy and the Environment and served on a number of advisory panels at national and international levels, including the Intergovernmental Panel on Climate Change (IPCC).

4   There are various definitions of the term "adaptive capacity." Here it is understood as "the ability of a system to adjust to climate change (including climate variability and extremes) to moderate potential damages, to take advantage of opportunities, or to cope with the consequences" (Gallopin 2006, 300).

References

Adger, W.N. 2003. "Social Capital, Collective Action, and Adaptation to Climate Change." *Economic Geography* 79, 4: 387–404.

Ahmed, A.K. 2006. "Concepts and Practices of 'Resilience': A Compilation from Various Secondary Sources." USAID/Asia Indian Ocean Tsunami Warning System Program and Coastal Community Resilience Program. http://www.adpc.net/v2007/programs/ews/CCR/downloads/CCRConceptsandPracticesofResilience.pdf.

Association of Municipalities of Ontario (AMO). 2008. *A Sustainability Planning Toolkit for Municipalities in Ontario.* Toronto: AMO.

Ayres, R., and E. Ayres. 2010. *Crossing the Energy Divide: Moving from Fossil Fuel Dependence to a Clean-Energy Future.* Saddle River, NJ: Pearson Education.

Balls, J. 2010. "Transition Towns: Local Networking for Global Sustainability?" Undergraduate diss., University of Cambridge.

Bay Localize. 2009. *Community Resilience Toolkit: A Workshop Guide for Community Resilience Planning.* Oakland: Bay Localize.

Cadell, S., J. Karabanow, and M. Sanchez. 2001. "Community, Empowerment, and Resilience: Paths to Wellness." *Canadian Journal of Community Mental Health* 20, 1: 21–35.

Caldwell, W.J. 2013. "Peak Oil and Climate Change: A Rural Community Guide." University of Guelph. http://www.waynecaldwell.ca.

Callaghan, E.G., and J. Colton. 2008. "Building Sustainable and Resilient Communities: A Balancing of Community Capital." *Environment, Development, and Sustainability* 10, 6: 931–42.

Canadian Institute of Planners (CIP). 2010. "Mainstreaming Climate Change Tools for the Professional Planning Community." http://www.planningforclimatechange.ca/.

Colussi, M. 2000. *The Community Resilience Manual: A Resource for Rural Recovery and Renewal.* Port Alberni, BC: Centre for Community Enterprise.

Colussi, M., M. Lewis, S. Lockhart, S. Perry, P. Rowcliffe, and D. McNair. 1999. "The Community Resilience Manual: A New Resource Will Link Rural Revitalization to CED Best Practice." *Making Waves* 10, 4: 10–14.

Community Research Connections. 2010. *Meeting the Climate Change Challenge.* Royal Rhodes University. http://www.crcresearch.org/climate-change/planning-adaptation-and-resiliency-canadian-local-government-experiences-and-needs.

Coutu, D. 2002. "How Resilience Works." *Harvard Business Review* 80, 5: 46–51.

Curtis, F. 2009. "Peak Globalization: Climate Change, Oil Depletion, and Global Trade." *Ecological Economics* 69, 2: 427–34.

Diamond, J.D. 2006. *Collapse: How Societies Choose to Fail or Succeed.* New York: Penguin.

Edwards, C. 2008. "Resilient Nation: Next Generation Resilience Relies on Citizens and Communities, Not the Institutions of State." http://www.sd-commission.org.uk/publications/downloads/Resilient_Nation.pdf.

Environmental Commissioner of Ontario. 2009. *Building Resilience Annual Report 2008–2009.* http://www.eco.on.ca/uploads/Reports-Annual/2008_09/ECO-Annual-Report-2008-2009.pdf.

Expert Panel on Climate Change Adaptation. 2009. *Adapting to Climate Change in Ontario.* http://www.ene.gov.on.ca/publications/7300e.pdf.

Friedrichs, J. 2010. "Global Energy Crunch: How Different Parts of the World Would React to a Peak Oil Scenario." *Energy Policy* 38, 8: 4562–69.

Gallopin, G.C. 2006. "Linkages between Vulnerability, Resilience, and Adaptive Capacity." *Global Environmental Change* 16: 293–303.

Gardner, S.M., and D. Noble, eds. 2008. *Stepping Up to the Climate Change Challenge: Perspectives on Local Government Leadership, Policy, and Practice in Canada.* Municipal Knowledge Series. St. Thomas, ON: Municipal World.

Gilbert, R. 2006. "Hamilton: The Electric City." Report written for the City of Hamilton.

Groundswell Group. 2007. *The Impact of Peak Oil on Rural Communities.* Cornwall, UK: Groundswell Group.

Grubb, A. 2010. *Peak Oil Primer.* http://www.energybulletin.net/primer.php.

Harvey, L.D.D. 2009. "No Time to Waste." *Canadian Dimension* 43, 6: 22–25.

Heinberg, R. 2007. *Peak Everything: Waking Up to the Century of Declines.* Gabriola, BC: New Society Publishers.

Holmgren, D. 2009. *Future Scenarios: How Communities Can Adapt to Peak Oil and Climate Change.* White River Junction, VT: Chelsea Green.

Homer-Dixon, T., and N. Garrison, eds. 2009. *Carbon Shift: How the Twin Crises of Oil Depletion and Climate Change Will Define the Future.* Toronto: Random House.

Hopkins, R. 2006. "Energy Descent Pathways: Evaluating Potential Responses to Peak Oil." MSc diss., University of Plymouth.

Jaccard, M. 2009. "Peak Oil and Market Feedbacks." In *Carbon Shift: How the Twin Crises of Oil Depletion and Climate Change Will Define the Future,* edited by T. Homer-Dixon and N. Garrison, 96–131. Toronto: Random House.

Keith, D. 2009. "Dangerous Abundance." In *Carbon Shift: How the Twin Crises of Oil Depletion and Climate Change Will Define the Future,* edited by T. Homer-Dixon and N. Garrison, 27-58. Toronto: Random House.

Klug, Jessica. 2009. "Addressing the Impacts of Peak Oil and Climate Change: An Analysis of the Ontario *Green Energy and Green Economy Act, 2009.*" Major paper presented to the Faculty of Graduate Studies, University of Guelph.

Kulig, J.C., D.S. Edge, and B. Joyce. 2008. "Understanding Community Resiliency in Rural Communities through Multimethod Research." *Journal of Rural and Community Development* 3, 3: 76–94.

Kunstler, J.H. 2006. *The Long Emergency: Surviving the End of Oil, Climate Change, and Other Converging Catastrophes of the Twenty-First Century.* New York: Grove Press.

Lerch, D. 2007. *Post Carbon Cities: Planning for Energy and Climate Uncertainty.* Sebastopol: Post Carbon Press.

——. 2009. "Energy Uncertainty and Community Resilience." http://www.postcarbon.org/article/40394-energy-uncertainty-and-community-resilience.

MacDonald, B. 2009. "Resilient Communities in Atlantic Canada: A Background Paper." Antigonish, NS: St. Francis Xavier University.

Magis, K. 2007. "Community Resilience Literature and Practice Review." US Roundtable on Sustainable Forests, September, Special Session on Indicator 38: Community Resilience. http://www.sustainableforests.net/summaries.php.

Magis, K. 2010. "Community Resilience: An Indicator of Social Sustainability." *Society and Natural Resources* 23: 401–16.

Marsden, W. 2009. "The Perfect Moment." In *Carbon Shift: How the Twin Crises of Oil Depletion and Climate Change Will Define the Future,* edited by T. Homer-Dixon and N. Garrison, 153–176. Toronto: Random House.

Maybery, D., R. Pope, G. Hodgins, Y. Hitchenor, and A. Sheperd. 2009. "Resilience and Well-Being of Small Inland Communities: Community Assets as Key Determinants." *Rural Society* 19, 4: 326–39.

Mehdi, B., C. Mrena, and A. Douglas. 2006. *Adapting to Climate Change: An Introduction for Canadian Municipalities.* http://www2.gnb.ca/content/dam/gnb/Departments/env/pdf/Climate-Climatiques/AdaptingClimateChangeIntroduction.pdf.

Natural Resources Canada (NRC). 2008. *From Impacts to Adaptation: Canada in a Changing Climate 2007—Synthesis.* Ottawa: NRC.

Nelson, D.R., W.N. Adger, and K. Brown. 2007. "Adaptation to Environmental Change: Contributions of a Resilience Framework." *Annual Review of Environment and Resources* 32, 1: 395–419.

Newman, P., T. Beatley, and H. Boyer. 2009. *Resilient Cities: Responding to Peak Oil and Climate Change.* Washington, DC: Island Press.

Noble, D., and A. Abram. 2008. "Overview: A New Look at the Big Picture." In *Stepping Up to the Climate Change Challenge: Perspectives on Local Government Leadership, Policy, and Practice in Canada,* edited by S.M. Gardner and D. Noble, 7–15. Municipal Knowledge Series. St. Thomas, ON: Municipal World.

Norris, F.H., S.P. Stevens, B. Pfefferbaum, K.F. Wyche, and R.L. Pfefferbaum. 2008. "Community Resilience as a Metaphor, Theory, Set of Capacities, and Strategy for Disaster Readiness." *American Journal of Community Psychology* 41, 1–2: 127–50.

Ontario. 2011. *Ontario's Adaptation Strategy and Action Plan 2011–2014*. Toronto: Queen's Printer for Ontario.

Owen, A. 2009. "What Makes Places Resilient? Are Resilient Places Sustainable Places?" http://www.sd-commission.org.uk/data/files/publications/resilience_of_places.pdf.

Partridge, M.D. 2007. "Rural Economic Development Prospects in a High Energy Cost Environment." *Journal of Regional Analysis and Policy, Special Issue on Rural Development Policy*, 37, 1: 44–47.

Peñuelas, J., and J. Carnicer. 2010. "Climate Change and Peak Oil: The Urgent Need for a Transition to a Non-Carbon-Emitting Society." *AMBIO: A Journal of the Human Environment*, 39 (1): 85–90.

Poland, B., A. Feitosa, C. Teelucksingh, D. Schugurensky, and R. Motzchnig-Pitrik. 2009. "Building Community Resilience: Mapping the Terrain and Refining the Practice." Project abstract.

Reid, S., B. Smit, S. Belliveau, and W. Caldwell. 2007. "Vulnerability and Adaptation to Climate Risks in Southwestern Ontario Farming Systems." In *Farming in a Changing Climate: Agricultural Adaptation in Canada*, edited by E. Wall, B. Smit, and J. Wandel, 173–86. Vancouver: UBC Press.

Reynolds, N. 2010. "There Is No 'Peak Oil.' But There Is Supply and Demand." *Globe and Mail*, 23 June, B2.

Richardson, G.R.A. 2010. *Adapting to Climate Change: An Introduction for Canadian Municipalities*. Ottawa: NRC.

Rowell, A. 2010. *Communities, Councils, and a Low-Carbon Future*. Totnes, UK: Green Books.

Standing Senate Committee on Agriculture and Forestry. 2003. *Climate Change: We Are at Risk. Final Report*. http://www.parl.gc.ca/37/2/parlbus/commbus/senate/com-e/agri-e/rep-e/repfinnov03-e.htm.

Torjman, S. 2007. *Shared Space: The Communities Agenda*. Ottawa: Caledon Institute for Social Policy.

Wall, E., B. Smit, and J. Wandel, eds. 2007. *Farming in a Changing Climate: Agricultural Adaptation in Canada*. Vancouver: UBC Press.

White, R. 2010. *Issues in Canada: Climate Change in Canada*. Don Mills, ON: Oxford University Press.

CHAPTER ONE

BUILDING COMMUNITY RESILIENCE IN HURON COUNTY: LESSONS FROM THE 2011 GODERICH TORNADO

SUSANNA REID

On 21 August 2011, at 3:53 p.m., a tornado hit the town of Goderich, Ontario. The scale used to measure the intensity of tornadoes is called the Fujita scale, which ranges from F0 (weakest at 110 kilometres per hour) to F5 (strongest at 450 kilometres per hour). Ninety percent of the tornadoes in Ontario register as F0 or F1 on the Fujita scale.

The tornado that hit Goderich was rated F3 by Environment Canada, with peak winds between 250 and 320 kilometres per hour. The last time that a tornado of this intensity was reported in Ontario was in 1996. Environment Canada was able to provide a fifteen-minute warning to the town before it hit (Rotteau 2011).

The path of the tornado was nineteen kilometres long. At the widest point, the path was 1.5 kilometres wide. The tornado was on the ground for fourteen seconds. People in town at the time said that it sounded like a train.

GODERICH, HURON COUNTY

Huron County is an agricultural region that had a population of 59,100 in 2011 (Statistics Canada 2011a). The Huron County population is characteristic of the population in rural Ontario and was stable or growing slightly from 52,951 in 1971 to 59,100 in 2011 (Statistics Canada 1972, 2011b). Since 1996, the Huron population has declined by 1.8 percent, from 60,220 in 1996 to 59,100 in 2011 (Statistics Canada 2001b).

The rural nature of Huron County means that 60 percent of its population live on a farm or rural property (Huron Business Development Centre 2010). Goderich is the largest town in Huron County, with a population of 7,521 in 2011 (Statistics Canada 2011a). Its population has been stable over the past several census periods (Statistics Canada 2001a). The Goderich population (median age of 48.5 years in 2011) is older than the Huron County population (median age of 46.0 years) and the provincial population (median age of 40.4 years) (Statistics Canada 2011a, 2011b).

Goderich, located on the shore of Lake Huron at the mouth of the Maitland River, is the county seat. The proximity to Lake Huron makes Goderich a popular tourist destination through the summer months. The Goderich harbour is the only deep-water commercial port on the east side of Lake Huron (Goderich Port Management Corporation 2013) and is home to the Sifto Salt Mine. Although manufacturing is important to the local economy, the sector suffered a significant blow in 2010 when Volvo Construction Equipment relocated a grader-manufacturing facility to the United States, with a loss of about 500 jobs.

The town boasts a unique street plan designed by John Galt when Goderich was founded in 1850 (Planning Partnership 2012). The downtown core, referred to as "The Square," is an octagon with Court House Park in the centre. Like most downtown areas in rural Ontario, The Square has been experiencing pressure from highway commercial development.

THE 2011 GODERICH TORNADO

The destruction left behind by the tornado was immense. In its wake, there was one fatality, and thirty-seven people were injured. Mature trees were uprooted and broken, hydro poles were snapped like matchsticks, hydro wires were snaking over the ground, people's homes were missing roofs. Walls were gaping open, leaving houses resembling life-sized dollhouses. Many buildings in the core area of town were damaged beyond repair. Everywhere there were piles of debris—trees and branches in full leaf, remains of buildings, walls, roofs, siding.

The Goderich Building Department (2012) issued thirty-eight residential demolition permits and twenty commercial demolition permits following the storm. The tornado's path is shown in Figure 1.6. As it continued past Goderich, the tornado damaged sixty-six properties and destroyed 200 acres of woodlot (Huron County Planning Department 2012).

The tornado was destructive not only in its force but also in its route. It directly hit three significant areas of employment: the Sifto Salt Mine at the harbour (488 jobs), the downtown core (839 full-time and part-time retail and office jobs), and an industrial area in the northwest end of town (approximately 250 jobs) (see Figure 1.7) (Armstrong 2013; Howe 2013).

The scientific community reports that there are not adequate data to indicate whether there has been an increasing occurrence of tornadoes as a result of climate change (Hoerling 2011; Solomon et al. 2007). Although tornadoes might not be characteristic of extreme weather events associated with climate change, they do represent the type of unpredictable event expected to become more frequent in the future. For this reason, the 2011 Goderich tornado is considered a case study to discuss the resilience of rural Ontario communities in anticipation of more frequent severe and unexpected events such as those associated with climate change and peak oil.

Figure 1.1. Looking east on West Street at the County Courthouse, Goderich

Figure 1.2. St. Patrick Street and Arthur Street, Goderich

Figure 1.3. Victoria Street United Church, Goderich

Figure 1.4. Damaged trees and hydro wires in Goderich

Figure 1.5. Surveying the damage after the tornado in Goderich

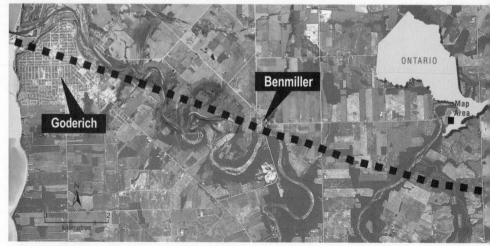

Figure 1.6: Path of 2011 tornado outside Goderich

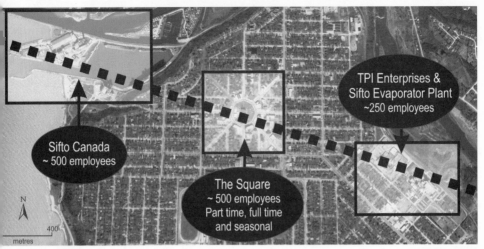

Sifto Canada
~ 500 employees

The Square
~ 500 employees
Part time, full time
and seasonal

TPI Enterprises &
Sifto Evaporator Plant
~250 employees

N

400
metres

Figure 1.7: Path of 2011 tornado within Goderich

Figure 1.8. Damaged buildings at Sifto Salt Mine, Goderich

Figure 1.9. Damaged buildings on West Street, Goderich

Figure 1.10. Damaged buildings in The Square, Goderich

Figure 1.11. Damaged buildings at the corner of Kingston Street and The Square, Goderich

Figure 1.12. Sifto Evaporator Plant, Goderich

This chapter offers frameworks for understanding transformation in socio-ecological systems and looks to the experience of the 2011 Goderich tornado to consider community resilience in response to an unpredicted catastrophe. Following the Goderich tornado case study, two community development efforts in Huron County are considered to provide more general observations about building community resilience in rural Ontario: the Huron Water Protection Steering Committee and Take Action for Sustainable Huron.

DEFINING RESILIENCE

Resilience means the ability of a system to withstand stress and recover to a similar state (Ferguson and Lapierre-Fortin 2012; Taleb 2012). The concept of resilience is found in many different areas of literature. It originated in the ecological literature but is an important idea in the public health, mental health, disaster response, socio-ecological, community development, natural resource management, and climate change literature (Gallopin 2006, in Ferguson and Lapierre-Fortin 2012). Resilience theory provides a dynamic, systems-oriented framework that is transferrable to any type of system—person, household, neighbourhood, business, farm, institution, forest, ecosystem, town, community, et cetera (Nelson, Adger, and Brown 2007). More discussion is provided in this chapter to conceptualize transformation and resilience in systems.

COMMUNITY RESILIENCE IN GODERICH

Adaptive capacity refers to the social and physical resources within a system that influence that system's ability to adapt. It is a core characteristic of a resilient socio-ecological system (Nelson, Adger, and Brown 2007). Wall and Marzall (2006) have provided a framework to characterize the resources that define the adaptive capacity of a community. The social resources defined in their framework are used to summarize the resources available to Goderich to respond to the 2011 tornado (see Table 1.1).

There is some overlap among the categories of resources in Goderich. For example, the capacity of the social network was evident in the friends, family, and community members who immediately responded to help people whose houses had been damaged as well as the area municipalities, social service agencies, and utility companies. The human resources in the utility companies, emergency services, and municipal administration played significant roles in the emergency response.

There was extensive capacity available to Goderich residents to respond to the tornado. The town was assisted by skilled regional and provincial teams of emergency responders and hundreds of volunteers who came to Goderich to assist in any way they could. Following the immediate efforts of the first

RESOURCE	DEFINITION	AGENCIES AND RESOURCES IN GODERICH FOLLOWING 2011 TORNADO
SOCIAL	People's relationships with each other through networks and associational life in their community	**SOCIAL SERVICE AGENCIES:** Red Cross United Way Ontario Society for the Prevention of Cruelty to Animals (OSPCA) Ontario 211 (community telephone helpline)
HUMAN	Skills, education, experiences, and general abilities of individuals combined with the availability of "productive" individuals	**MUNICIPAL ADMINISTRATION:** Town of Goderich Neighbouring municipalities Huron County **EMERGENCY SERVICES:** Alexander Marine General Hospital Ontario Provincial Police (eighty officers) Six area fire departments Huron Emergency Medical Services **UTILITIES:** Union Gas Eight neighbouring hydro companies
INSTITUTIONAL	Government-related infrastructure (fixed assets)—utilities such as electricity, transportation, water, institutional buildings, and services related to health, social support, and communication	**PROVINCIAL AGENCIES:** Ministry of Labour Critical Incident Stress Management **COMMUNITY RESOURCES:** Knights of Columbus Community Centre Faith communities 1,200 people who phoned the town to offer assistance
ECONOMIC	Financial assets, including built infrastructure as well as a number of features enabling economic development	Insurance payments approximately $110 million Ontario Disaster Relief Assistance Program Local fundraising: $4 million Provincial contribution: $2.1 million Edgefund (funds community responses to the tornado) $170,000

Table 1.1. Adaptive capacity of Goderich, Ontario

couple of weeks, insurance payments and local and provincial assistance programs financed the reconstruction of damaged residences and businesses.

A FRAMEWORK FOR UNDERSTANDING TRANSFORMATION IN SYSTEMS

Gunderson and Holling (2002) have developed a framework called "panarchy" for understanding transformation in human and natural systems (Figure 1.13).

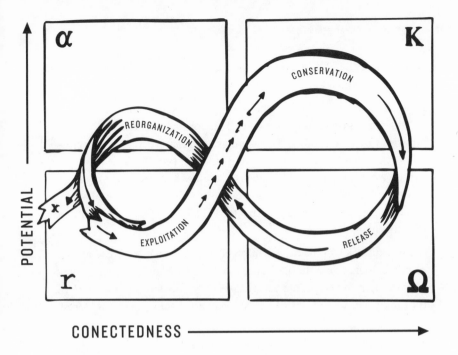

ure 1.13. Panarchy model

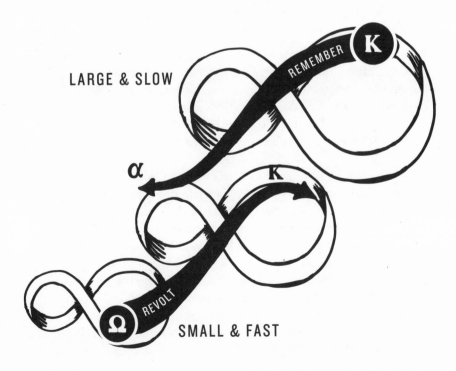

gure 1.14. Nested panarchy cycles

As described by the panarchy model, any system includes four phases (Francis 2010; Gunderson and Holling 2002).

Exploitation (r to K). The first phase is a growth phase. Imagine a meadow that is becoming a forest. There is constant growth, and shrubs grow into grassy meadows, then saplings, then mature trees. The exploitation phase tends to occur over a long period of time. Because it takes time, the exploitation or growth phase becomes familiar. The growth phase tends to be perceived as good or positive. Because it is familiar, there tends to be an expectation that the exploitation phase will continue without change.

Conservation (K). In the conservation phase, the system uses available resources to maintain itself. It is not too different from the exploitation phase and, from an external perspective, might not look too different. In the example of a forest, the conservation phase would be a mature forest.

Release (K to omega). This phase represents a release of energy that enters the system into a phase of creative destruction. In the example of a forest, this could be a forest fire, a disease, or an insect infestation. The release phase is generally unexpected, unpredictable, and chaotic. It can be stressful for agents who are part of the system or influenced by it. It is also difficult to determine how to respond and manage change during the release phase.

Reorganization (omega to alpha). The final phase of the panarchy model is reorganization. This is the inevitable result of the release phase. After the system has entered into creative destruction, it must reorganize itself. The reorganization phase does not presume that the system will return to its initial form. It might look completely different from the original expression. It is during this phase that novelty and innovation can occur.

The word *panarchy* was coined by Gunderson and Holling (2002) by combining a word and a name: the word *hierarchy*, referring to nested systems, and the name of the Greek god Pan, the goat-legged god of wine and dance. Figure 1.14 provides an illustration of the hierarchical nature of the panarchy model.

This illustration suggests that there is a hierarchical relationship among systems (as suggested by the name panarchy). Gunderson and Holling refer to "nested systems." This hierarchical model also demonstrates the importance of scale (Nelson, Adger, and Brown 2007; Wall and Marzall 2006).

In the example of the Goderich tornado, the system at the bottom can be a household, above that a neighbourhood, and above that the town of Goderich. One can imagine that, for most systems, many more panarchies can be added.

For example, they can be added to represent all of the households in Goderich and all of the neighbourhoods. There can be additional systems added above the top—"large and slow"—system, representing southwestern Ontario, and one above that representing the province of Ontario.

As shown in Figure 1.14, systems interact with and influence each other. The ability of individual Goderich households to respond to the impact of the tornado influenced the responsiveness of neighbourhoods and likewise the responsiveness of Goderich as a town. The reciprocal relationship is also true. The support of the province, the region of southwestern Ontario, and Huron County influenced the ability of Goderich, including the town itself and all of the neighbourhoods and individual households in the town, to adapt to the tornado.

The interconnectivity of hierarchical systems in the panarchy model demonstrates the complexity of communities. Many factors influence and contribute to change within communities. As characterized by the goat-legged god Pan, community change is dynamic and non-linear.

THE ROLE OF PLANNING

Community practitioners and planners have a significant opportunity to assist communities following an occurrence that represents a release of energy that moves a community (or part of a community) from the conservation phase to the release phase. This can be a community change, such as the loss of an employer, or some expression of community dissonance, such as an environmental concern. According to the panarchy model, something causes a release of energy and launches the system into a creative destruction phase. It is generally characterized as a stressful and unpredictable time.

The 2011 Goderich tornado is considered an example of a release of energy that sent the community of Goderich into a creative destruction/release phase. There were many community and public processes to assist Goderich in recovering from the tornado. Some examples are provided below.

- The Public Meeting and Community Support Forum, 27 August 2011, provided information and outlined community supports available.

- The Public Meeting and Community Workshop Regarding Parks and Public Spaces, 24 September 2011, invited the community to discuss and contribute ideas to redevelopment of the parks.

- The Goderich Master Plan, developed by Planning Partnerships, was intended to redesign and redevelop The Square in the downtown core.

- The Goderich Official Plan, Heritage Committee, and land-use planning tools provided direction for redevelopment of the core area.

The successful redesign and redevelopment of the core area of Goderich are testament to the effectiveness of community planning processes and tools.

ANTI-FRAGILITY

In a recent book titled *AntiFragile: Things that Gain from Disorder* (2012), Nassim Taleb provides breadth to the concept of resilience. Taleb suggests that there are three types of systems: (1) fragile systems that break when they are stressed, (2) resilient systems that can recover when stressed, and (3) anti-fragile systems that get stronger when stressed. In coining the term, Taleb suggests that there is no prior word in any language that names the concept of systems becoming stronger as a result of being stressed. The anti-fragility concept echoes Friedrich Nietzsche's famous colloquialism "what doesn't kill you makes you stronger." Communities, individuals, businesses, companies, and institutions all have opportunities to become stronger in response to setbacks or stresses.

Taleb's (2012) theory purports that systems become stronger when they respond to a stress sooner rather than later. Here is one analogy from Taleb to illustrate this point. A glass full of water can withstand being dropped an inch onto a table repeatedly without any water spilling. If the glass is dropped from waist height, it will spill water but might not break. Dropped from the height of a building, it will certainly break. Taleb's point is that, if a system responds to an observable stress when it is first observed, and adapts with the intention of strengthening the system, then there is a greater chance of success. If a system ignores or postpones responding to a stress, and the stress becomes greater, then the chance of the system being able to adapt is reduced since the stress might be too great (as with the glass of water analogy, the greater the stress, the more fragile the system).

Coupled with the panarchy theory, this theory suggests that a more responsive system will have a shorter recovery period after a stress is introduced and a less acute creative destruction phase.

GODERICH RESPONSES TO THE TORNADO

Examples of fragile, resilient, and anti-fragile systems can be found in the aftermath of the 2011 tornado.

Fragile Systems – Trees represent a fragile system unable to recover from the tornado. The Parks Department manager of Goderich reported 500 trees damaged in public parks (Rotteau 2011). Many more trees on private land were destroyed in Goderich. Outside the town, 200 acres of mature forests were destroyed (Huron County Planning Department 2012).

Resilient Systems – Many households and businesses damaged or destroyed by the tornado were repaired or rebuilt to similar states. This required significant resources and abilities, including adequate insurance coverage, project management abilities, and local tradespeople.

Anti-Fragile Systems – Residents of Goderich have a strong affinity with the downtown core. The original plan for the town is a unique design, with the County Courthouse in the centre of an octagon, eight radiating streets from the octagon, and four boundary streets creating a square around the centre. The octagon in the centre is named Courthouse Square and referred to as The Square.

The Square was well used prior to the tornado. Downtown festivals and the Saturday market were popular events. Although The Square was an important commercial destination and public space, it was a bit tired. There were a few closed stores, and some of the storefronts were due for upgrades. There had been gradual die-back of the mature trees in Courthouse Park over the years.

After the tornado, there was resounding community interest in the redesign and redevelopment of The Square. It seems that the tornado's destruction ignited a renewed passion for the downtown. The town engaged a planning consulting firm that hosted a three-day design charette and several public consultation sessions. Hundreds of people participated in the process with much enthusiasm and excitement for the potential of the public space (Boa 2012; Planning Partnership 2012; see Figure 1.22).

As of fall 2013, the town was well into implementation of the community's design. One hundred and fifty trees have been planted in the centre of The Square. New light fixtures and sidewalks have been installed. The Saturday farmers' market has been relocated to be closer to the retail stores. New gardens have been installed. Together with the redevelopment and reconstruction of several stores that were damaged, the town centre has a new sense of optimism. The community's relationship to The Square following the tornado represents an anti-fragile system that has become stronger as a result of being stressed.

ANTI-FRAGILITY, PANARCHY, AND COMMUNITY DEVELOPMENT

Wedded with the panarchy model, the anti-fragility concept suggests that, when there is a release of energy and community change occurs, there is a critical opportunity for community development practitioners to assist with reorganization or redevelopment. If done effectively, this can result in stronger, more resilient communities.

The role of community practitioners has three phases.

Recognize the change in the community, expression of dissonance, or release of energy.

Design and host a public process or plan to recognize the significance of the change and, through public discussion, support the community through the creative destruction phase.

Work to implement the plan once it is developed. This can be a dynamic process, involving public, business, and non-governmental stakeholders. There is a need to secure financing and resources and for project development, implementation, and management.

BUILDING COMMUNITY RESILIENCE IN HURON COUNTY

Although most changes in communities are not as catastrophic as the Goderich tornado, the panarchy model suggests that community practitioners can expect change and actively seek opportunities to be involved in strengthening communities. Often community transformation is expressed as community conflict or dissonance. A consultative community process, if provided, can allow people to contribute more productively.

Two examples of Huron County processes to assist with building community resilience are reviewed below: Take Action for Sustainable Huron and the Huron County Water Protection Steering Committee. These are examples of efforts to develop community resilience in less extreme conditions than recovering from the Goderich tornado.

TAKE ACTION FOR SUSTAINABLE HURON

Take Action for Sustainable Huron was adopted by Huron County Council in December 2011, building on two years of community consultation and recognition of the strengths and assets of the region (Huron County 2011). The Take Action plan considers environmental, social, cultural, and economic goals in a holistic fashion to chart future directions for community development. Development of the plan is itself an effort to build community resilience. Two of many project areas developing under the auspices of the Sustainable Huron plan are local food and transportation.

Local Food – Agriculture is a pillar of the Huron County economy. There are 2,467 farms in Huron County, with a total annual farm income of $765 million in 2011 (OMAF 2013). Demands of the international market require increased

igure 1.15. St. David Street, Goderich

Figure 1.16. Force of tornado splinters mature trees

Figure 1.17. November 2012 parade of trees in The Square, Goderich

Figure 1.18. House on St. Patrick Street, Goderich, after tornado

Figure 1.19. Same St. Patrick Street property after the tornado, showing the reconstructed house

Figure 1.20. The Square in Goderich before the 2011 tornado

Figure 1.21. The Square in Goderich immediately after the 2011 tornado

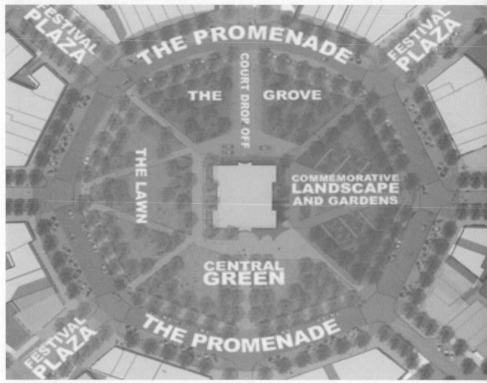

Figure 1.22. Master plan for The Square, Goderich

Figure 1.23. Looking west, the damaged tree canopy on the north side marks the tornado's path

efficiencies, which have resulted in farm consolidations. The total number of farms in Huron County declined by 413 between 2001 and 2011, from 2,880 in 2001 to 2,467 in 2011 (ibid.). The greatest loss of farm numbers has been mid-sized farms (between 130 and 400 acres). Another change in the Huron County food sector has been a loss in agricultural processing jobs. Cangro in Exeter closed in 2008 (*Toronto Star* 2008), with a loss of 130 jobs, and E.D. Smith in Seaforth closed in 2013, with a loss of 180 jobs (Murray 2012).

There are several community development efforts under way to build resilience in the agri-food sector through supporting local growing, dining, retailing, wholesaling, and processing. Efforts include the following.

▶ There is a new Bruce-Huron Wholesale Auction in Lucknow, where growers within a seventy-five-kilometre radius of Lucknow are selling wholesale quantities of vegetables and fruits. The auction is entering its third year. As of 2014, the auction has been open for four years.

▶ The Huron Perth Farm to Table map generates farm gate sales for local producers (Huron Perth Farm to Table 2013).

▶ The award-winning Taste of Huron entered its fifth year in 2013, celebrating local food and building a stronger relationship between local growers and restaurants (Taste of Huron 2013).

▶ Huron County Economic Development Services is exploring business scenarios for the development of a viticulture industry (Huron Economic Development Services and Huron Business Development Centre 2012).

Transportation – Huron County is characterized by a low-density population, with long distances between places. This creates logistical and financial challenges for providing a public transportation system. In Huron County, 80 percent of trips made are in single-occupancy vehicles (MMM Group 2011). The reliance on cars creates challenges for people who are disadvantaged by way of health, employment, or family circumstances (Social Research and Planning Council 2012). Distance makes it difficult to use active transportation (walking, cycling) as a regular mode of transit for most people.

In 2010, Huron County commissioned a Transportation Demand Management (TDM) Plan (MMM Group 2011). The plan identified several transportation strategies to improve active transportation and provide transportation options for the Huron County population. As a result of the Take Action plan, two initiatives are under way.

▶ The Huron Perth United Way is surveying social service organizations that provide transportation to different populations (e.g., Children's Aid Services has volunteer drivers, EasyRide provides transportation to medical appointments, etc.). A comprehensive review of the transportation services offered might lead to efficiencies in service delivery.

▶ An active transportation network has been created to develop and improve the existing network of walking, hiking, and cycling trails in the area.

Both of these efforts seek to build community resilience. Improving transportation services for vulnerable populations seeks to be more inclusive with service delivery. Enhancing an active transportation network will further develop an attractive tourism destination and improve recreational options for Huron County residents.

THE WATER PROTECTION STEERING COMMITTEE

In the early 2000s, there was growing tension between lakeshore residents and the agricultural community about water quality in near-shore Lake Huron. Local and provincial media headlines cited concerns about E. coli levels in the lake (Hall 2004; Spears 2003a, 2003b). Attempts to isolate the cause to a single source resulted in finger-pointing between the non-farm and farm communities and increasing levels of conflict.

Through the lens of the panarchy framework, the conflict over water quality can be considered a release of energy. If not addressed by the local government, then the tension between the lakeshore and agricultural communities might have become more heated.

Instead, Huron County Council responded by establishing the Huron Water Protection Steering Committee (WPSC) in the spring of 2004. The WPSC has three mandated goals: (1) to bring together representatives of agencies, groups, and municipalities, (2) to prioritize and recommend implementation measures to participating agencies, and (3) to coordinate activities at a broad level, subject to the resources of the participating agencies (Caldwell and Tousaw 2004).

By establishing the WPSC, Huron County Council recognized that, while many agencies and organizations have mandates to address water quality, there was no coordinating body. The WPSC has initiated conversations between farmers and lakeshore residents, shared information about watershed management, coordinated funding applications, and fostered a collaborative approach to the common concern over water quality.

Creation of the WPSC has assisted in efforts to "reorganize" (as suggested by the panarchy model) the system of watershed management in Huron County. Sharing responsibility for resource management builds resilience through enhanced efficiency of decision making, increases levels of trust between landowner communities and the local government, and stretches available resources (Ferguson and Lapierre-Fortin 2012; Nelson, Adger, and Brown 2007).

There have been many projects developed and discussed at the Huron Water Protection Steering Committee. Two examples are provided here.

The Huron Clean Water Project – The Huron Clean Water Project is a collaborative effort financed by Huron County and delivered through the local stewardship efforts of Ausable Bayfield Conservation Authority and Maitland Valley Conservation Authority. From 2004 to 2012, Huron County contributed $2.27 million to a total project value of $5.8 million (Hocking 2012).

The project funds 50 percent of the cost of stewardship projects on private land. Eligible grant categories include tree planting on fragile land, fencing cattle out of waterways, decommissioning manure storage, decommissioning abandoned wells, upgrading wells, and undertaking erosion control.

The project is considered both an environmental effort to improve land stewardship and economic development work. For every dollar contributed by Huron County to stewardship efforts, 2.75 dollars are contributed by landowners or other funders to implement the project. By acting locally, the community is investing in water quality improvement, erosion control, and soil enhancement.

Healthy Lake Huron – The federal government (Environment Canada) and the provincial government (Ministries of Agriculture and Food, Environment, and Natural Resources) have launched an effort to develop a collaborative approach to addressing Lake Huron water quality. Healthy Lake Huron brings together agencies with jurisdictions in the Lake Huron watershed to develop a coordinated approach to addressing Lake Huron water quality.

Healthy Lake Huron projects include developing a rural storm water management plan and developing and implementing watershed action plans in five Lake Huron watersheds (Healthy Lake Huron 2013). Similar to the WPSC, through the development of a collaborative approach to water management, Healthy Lake Huron is building resilience.

CONCLUDING THOUGHTS

Building resilience in rural Ontario requires an inherent respect for the dynamic nature of communities as systems. Change will always occur. Within the complex, non-linear, goat-legged dance of community transformation, we can work together to build capacity for an uncertain future. The following are key points for building community resilience.

► Building resilience (or anti-fragility) is creative. It is a creative exercise to design community processes and broadly welcome and engage people in discussing what could happen. Community engagement requires an open-minded approach to understanding complex systems.

► It is crucial to recognize systems that are stressed or vulnerable and help them to reorganize. Changes in employment situations, such as the announcement of a plant closure, or widespread concern over an environmental condition such as water quality are examples of community-wide events that can represent a "release" of energy and precipitate community "reorganization." Community practitioners and agency staff have important roles to play in recognizing community-level change or dissonance and responding with opportunities for community dialogue, vision, projects, and actions.

► Partnerships are easier to build around projects. Sometimes community conflict is precipitated because of cultural differences. For example, the residents along the shore of Lake Huron and the neighbouring agricultural community are culturally very different from each other. A project such as a watershed plan that broadly engages all of the landowners in a watershed provides a focus for initiating conversations and building relationships.

► Thinking about how people make a living is important. In rural Ontario, many households are supported by resource-based economies. Successful community development efforts must be respectful of livelihoods. For example, environmental efforts in agricultural communities will be more successful if they support and enhance farming practices.

► Recognize contributions and celebrate accomplishments. When people respond to invitations to participate in community events, they are contributing generously to the community's future. If this expression is recognized and celebrated, then they will be inclined to invest again and share their experiences with others. Welcoming everyone is very important.

► Leadership is necessary (and everyone is a leader). Building community resilience (and anti-fragility) requires the strongest contribution of every member of a community. It requires that all citizens share their gifts and strengths and that they be recognized for doing so. It requires that we all act intentionally for the public good.

References

Armstrong, S. 2013. Personal communication with Susan Armstrong, Coordinator, Goderich Business Improvement Association, 13 May.

Boa, H. 2012. "Master Plan Creates Dynamic Downtown in Goderich." *Huron News Now,* 2 May. http://huron.bulletnewscanada.ca/2012/05/02/master-plan-creates-dynamic-downtown-in-goderich/.

Caldwell, W., and S. Tousaw. 2004. "Establishing a Water Protection Steering Committee." Report to Huron County Planning Agriculture and Public Works Committee, Goderich, ON, 2 January.

Ferguson, E., and E. Lapierre-Fortin. 2012. *Peak Oil, Climate Change, and Rural Ontario Community Resilience.* http://www.waynecaldwell.ca.

Francis, G. 2010. "The Hardcore Guide to Resilience." *Alternatives Journal: Environmental Ideas and Action* 36, 2: 13–15.

Gallopin, G.C. 2006. "Linkages between Vulnerability, Resilience, and Adaptive Capacity." *Global Environmental Change* 16: 293–303.

Goderich Building Department. 2012. "Tornado Demolition Permits Issued." Goderich, ON, September.

Goderich Port Management Corporation. 2013. Website. http://www.goderichport.ca/Port_Corporation/.

Gunderson, L.H., and C.S. Holling. 2002. *Panarchy: Understanding Transformation in Human and Natural Systems.* Washington, DC: Island Press.

Hall, J. 2004. "What Is Fouling the Beaches of Huron? E. Coli Source Subject of Angry Debate." *Toronto Star,* 1 March.

Healthy Lake Huron. 2013. Website. http://www.healthylakehuron.ca.

Hocking, D. 2012. "2012 Clean Water Project Summary to Huron County Committee of the Whole Day 1," Goderich, ON, April.

Hoerling, M. 2011. *Preliminary Assessment of Climate Factors Contributing to the Extreme 2011 Tornadoes.* National Oceanic and Atmospheric Administration, US Department of Commerce. http://www.esrl.noaa.gov/psd/csi/events/2011/tornadoes/reference.html.

Howe, R. 2013. Email correspondence from Rowland Howe, Director of Corporate Engineering, Strategic Projects, Compass Minerals, 13 May.

Huron Business Development Centre. 2010. *State of the Huron County Economy: Towards Sustainable Economic Renewal.* http://www.smallbusinesshuron.ca/publications.htm.

Huron County. 2011. "Take Action for Sustainable Huron: Community Sustainability Plan." Goderich, ON, December.

Huron County Planning Department. 2011. "Goderich Tornado Volunteer List," Goderich, ON.

——. 2012. "2011 Tornado Path beyond Goderich," Goderich, ON.

Huron Economic Development Services and Huron Business Development Centre. 2012. *Viticulture in Huron County: An Introduction for Investors.* http://www.investinhuron.ca/investment-opportunites/current-initiatives/.

Huron Perth Farm to Table. 2013. Website. http://www.huronperthfarmtotable.ca.

Jewell, D. 2013. Email correspondence from D. Jewell, Chair, Goderich ODRAP Committee, 4 February.

MMM Group. 2011. *Transportation Demand Management Plan Final Report.* Huron County. http://www.huroncounty.ca/sustainablehuron/tdmplan.php.

Murray, M. 2012. "Seaforth E.D. Smith Plant Closing Next Year." *Seaforth Huron Expositor,* 13 August. http://www.seaforthhuronexpositor.com/2012/08/13/seaforth-ed-smith-plant-closing-next-year.

Nelson, D., N. Adger, and K. Brown. 2007. "Adaptation to Environmental Change: Contributions to a Resilience Framework." *Annual Review of Environmental Resources* 32: 395–419.

Nichol, P. 2013. Email correspondence from Paul Nichol, Huron Business Development Centre, 5 February.

Ontario Ministry of Agriculture and Food (OMAF). 2013. *Huron County at a Glance.* http://www.omafra.gov.on.ca/english/stats/county/index.html.

Planning Partnership. 2012. *Town of Goderich Downtown Core Master Plan.* http://www.goderich.ca/en/townhall/Planning.asp.

Rotteau, L. 2011. "Goderich Tornado Report for EMO," Goderich, ON, 1 December.

Solomon, S., D. Qin, M. Manning, Z. Chen, M. Marquis, K.B. Averyt, M. Tignor, and H.L. Miller, eds. 2007. *Contribution of Working Group I to the Fourth Assessment Report of the Intergovernmental Panel on Climate Change.* Cambridge, UK: Cambridge University Press. http://www.ipcc.ch/publications_and_data/ar4/wg1/en/ch3s3-8-5.html.

Spears, T. 2003a. "Farmland and Septic Systems Undoing Decades of Great Lakes Clean-Up." *Ottawa Citizen,* 16 November.

——. 2003b. "'Ontario's West Coast' Permanently Polluted." *Ottawa Citizen,* 15 November.

Statistics Canada. 1972. *1971: Population Census Divisions and Subdivisions.* Ottawa: Statistics Canada.

——. 2001a. *Goderich, Ontario 2001 Community Profile.* http://www12.statcan.ca/.

——. 2001b. *Huron County, Ontario 2001 Community Profile.* http://www12.statcan.ca/.

——. 2011a. *2011 Statistics Canada Census Profile Goderich, Ontario.* http://www12.statcan.gc.ca/.

——. 2011b. *2011 Census of Canada: Topic Based Tabulations Ontario.* http://www12.statcan.gc.ca/.

The Sustainable Scale Project. 2013. Website. http://www.sustainablescale.org/ConceptualFramework/UnderstandingScale/MeasuringScale/Panarchy.aspx.

Taleb, N. 2012. *Antifragile: Things that Gain from Disorder.* New York: Random House.

Taste of Huron. 2013. Website. http://www.tasteofhuron.ca.

Toronto Star. 2008. "Cangro to Shut Exeter Plant, Eliminate 130 Jobs." *Toronto Star,* 7 April, B2.

Tremblay, R. 2013. Email correspondence from Robert Tremblay, Insurance Bureau of Canada, 7 February.

Wall, E., and K. Marzall. 2006. "Adaptive Capacity for Climate Change in Canadian Rural Communities." *Local Environment* 11, 4: 373–97.

CHAPTER TWO

POLICY AND GREEN INFRASTRUCTURE: PLANNING IN RESPONSE TO CLIMATE CHANGE AND PEAK OIL

PAUL KRAEHLING AND WAYNE CALDWELL

INTRODUCTION

The role of intact natural systems in resiliency work is critical to the long-term health of communities and in dealing with difficult future challenges such as climate change and diminished fossil fuels (peak oil). The interactions of community members and governments in acknowledging, respecting, and using the goods and services of nature are critical for planning resilient communities. This chapter argues that communities that have worked alongside the natural assets bestowed on their areas will be in better positions to respond and adapt to future unknowns. Mechanisms for environmental conservation, biodiversity enhancement, and stewardship of land, water, and air resources are discussed within an understanding of the need to conserve the land base in a healthy ecosystem manner.

We make an association in this chapter between the maintenance (or creation) of a healthy land base as a first condition of addressing the challenges of climate change and peak oil. We argue that the multifunctional benefits embedded in the goods and services of nature can assist in, and are necessary for, the planning and development of resilient communities. Because our background is in land-use planning, this general perspective informs our discussion (i.e., how can we plan using nature as the "keystone" to an overall sustainability perspective of balancing economic, social, and environmental factors for community development?).

We believe that the "carrying capacity" of the land base, as a general philosophical element of community planning, needs to be acknowledged and that the coexistence of nature with human habitation must be respected.[1] The conservation of existing resources is an important tenet of community planning (Hodge and Gordon 2014). Conservation is the most efficient and cost-effective mechanism to plan for future uncertainties. The conservation mindset is an important element of protecting existing assets for future use in ways that might not be well understood today; planning for the long term with stewardship responsibilities today gives consideration to future generations not yet born.

The philosophical perspective of this chapter reflects the work of Aldo Leopold in *A Sand County Almanac: With Other Essays on Conservation from Round River* (1949). His main premise is that the land requires a basic health regime to support human health (Carrick 2012). In advocating for the health of the land, his most famous quotation endures: "A thing is right when it tends to preserve the integrity, stability and beauty of the biotic community. It is wrong when it tends otherwise" (Leopold 1949, 224–25).

The land base in rural areas of Ontario provides both the greatest opportunities for and the challenges to addressing the changes that might result from climate change and peak oil. The land base provides the foundation for natural systems that provide life for all through interactions of the atmosphere, lithosphere, hydrosphere, and biosphere. However, the geography of rural Ontario, with low population densities and large physical distances, also requires that human activity and enterprise use fossil fuels to move about and derive economic benefits from the landscape. Movement systems are primarily enabled by the motor vehicle, which allows rural residents to move about in such a landscape—between places of residence, employment, education, and leisure. The large resource extraction base, consisting of industrial farm operations in southwestern Ontario, also requires fossil fuels to transport goods and commodities and provide inputs to farm crop production (fertilizers, pesticides).

To manage the opportunities and constraints of the land for human activity is the primary mission of sustainable planning and development. This chapter extols the virtues of a conservation perspective in using the land base and emphasizes maintaining a healthy ecosystem as the basis for healthy rural communities. The future challenges to land use from climate change and peak oil will require the land to be resilient to change—and as the land is resilient, so will be rural communities.

Our discussion covers a diverse range of topics, all dealing with nature—its maintenance, protection, and possible enhancement—with connections made to mechanisms to build community resilience. The chapter outlines a planning approach to the use of nature, natural systems, open spaces, and green infrastructure to derive maximum utility in designing sustainable and resilient rural communities. The following perspectives are investigated in turn: (1) from a philosophical perspective in which examples in the world today illustrate how planning for healthy rural communities begins with planning for a robust natural support ecosystem; (2) from the perspective of the provincial senior level of government responsible for land-use planning in Ontario.

The next chapter continues dialogue on the theme of natural systems for resilient rural community development. However, it delves more into case studies that describe specific action mechanisms at a community level to describe the virtues of planning from a natural systems perspective.

GENERAL PHILOSOPHY: USE OF NATURE/NATURAL SYSTEMS
TO PLAN FOR RESILIENT RURAL COMMUNITIES

The importance of using green space in community planning has been recognized from at least the late nineteenth century. Hodge and Gordon, in *Planning Canadian Communities: An Introduction to the Principles, Practice, and Participants* (2014), outline conditions that promoted the advent of Garden Cities and the City Beautiful movement in the early 1900s. The use of open green areas in land-use planning is not a new concept. The Garden Cities notion by Ebenezer Howard in the late nineteenth century and other design principles used in early Canadian cities incorporated these areas as important components of healthy communities. It can be argued that proximity to green space—whether in urban or in rural areas—is an important component of basic community design.

Several noted landscape architects made the association between healthy living conditions and the provision of open spaces and parks. Thomas Adams, the father of Canadian planning, was hired by the Canadian government in the early twentieth century to prepare a report for the Canadian Conservation Commission (Caldwell 2011). Adams's 1917 book *Rural Planning and Development in Canada* was very influential in noting the challenges of rural life at the time. Other important documents associating the health of nature with community design and development in the twentieth century include *A Sand County Almanac* (Leopold 1949) and *Design with Nature* (McHarg 1969).

At the turn of the twenty-first century, several community designers advocated for green space to be used in a much more holistic sense, going beyond the "simple" provision of this space as a respite park area within the community's built-up area (Daniels and Daniels 2003; Randolph 2012; Rouse and Bunster-Ossa 2013). In addition, some planners have argued that green natural heritage areas, deemed to be "development constraint" lands, should also be nested in a holistic planning system. Arguments are currently made that all forms of valuable "green space" should be acknowledged and used in a central design construct for communities. This design framework is entitled green infrastructure (GI).

Several recent documents outlined in the following paragraphs advocate for the provision of GI to be used in promoting healthy natural conditions as well as the health and well-being of humans. GI embodies the natural built elements of the landscape as well as human-inspired natural functioning facilities. Many elements of health and wellness are embodied within the requirements for clean water, air, and land. Nature provides a cost-effective means (because of the externalities associated with the goods and services embedded in nature) of helping to provide these essential elements of life.

The following is a succinct definition of GI:

Green infrastructure is defined as natural vegetation and green technologies that collectively provide society with a broad array of products and services for healthy living. Green infrastructure takes many forms including but not limited to the following: ... forests, natural areas, greenways, streams/water bodies and riparian zones, meadows and agricultural lands; green roofs and green walls; parks, gardens and landscaped areas, community gardens, and other green open spaces; rain gardens, bioswales, engineered wetlands and storm water ponds.

Green infrastructure also includes soil—in volumes and qualities adequate to sustain living green infrastructure and absorb water, as well as technologies like porous paving, rain barrels, cisterns and structural soil. (Green Infrastructure Ontario Coalition 2012, 2, with slight word modifications.)

In considering specific applications of green infrastructure, the following are indicative of the mechanisms and circumstances in which the embedded goods and services of nature can be used for health and wellness conditions of humans as well as to retain natural ecosystem health and well-being. These services and functions include the protection of surface and groundwater quality, the provision of water supply, groundwater recharge, increased health opportunities through outdoor recreation, improved air quality, reduction of summer heat island effects, reduced flooding, local agricultural production (urban and rural), improved community cohesiveness, reduction in traditional "grey infrastructure" servicing needs (storm sewers, impervious surfaces), opportunities for community education, community amenity and attractiveness, reduced energy need, improved wildlife habitat, reduced noise pollution, provision of materials (where available) such as lumber and aggregates, and climate change mitigation (Center for Neighborhood Technology 2010).

The alignment between healthy ecosystems and healthy human settlements, in conjunction with the multiplicity of functions embedded in natural GI systems, is illustrated in Figure 2.1.

Figure 2.1 outlines the various general components of a GI planning framework that supports both human settlements and the natural environment. The left side of the figure outlines goods and services that nature provides to humans, such as economic, social, and community/physical/psychological health; the right side of the figure illustrates how the natural system supports itself through GI elements, ecosystem services and functions, and overall ecosystem health and biodiversity.

Several publications discuss the theme of GI as a basis for community design. In the United States, GI is discussed in instances that cover macro-scale

HUMAN SETTLEMENTS

NATURAL ENVIRONMENT AND ECOSYSTEM

SOCIO-ECONOMIC HEALTH

Income and Employment
Living and Working Conditions
Education and Lifestyle
Access to Services & Housing

COMMUNITY HEALTH

Sense of Community Identity
Social Capital
Community Empowerment
Culture

PHYSICAL HEALTH

Cardiovascular
Nervous System
Digestion
Endocrine Functions and Immunity
Respiratory
Bone Tissue

PSYCHOLOGICAL HEALTH

Relaxation from Stress
Cognitive Capacity
Positive Emotions
Attention Capacity

GREEN INFRASTRUCTURE ELEMENTS (ALL INTERRELATED)

Natural Plant/Animal Corridors
Countryside, Active and Idle
Housing Green Spaces and Gardens
Water Areas (Lakes, Rivers, Wetlands)
Urban Parks and Open Spaces
Forested/Treed Areas

ECOSYSTEM SERVICES AND FUNCTIONS

Air Purifications
Soil and Nutrient Cycling
Habitat Provision
Aesthetic and Spiritual
Climate and Radiation Regulation
Water Purification
Waste Recycling/Decompositions
Noise Pollution Control

ECOSYSTEM HEALTH

Air Quality
Energy and Material Cycling
Habitat and Species Diversity
Soil Structure
Water Quality
Ecosystem Resilience

Figure 2.1. Conceptual Framework: Green infrastructure and planning for sustainability. Adapted from K. Tzoulas et al., "Promoting ecosystem and human health in urban areas using green infrastructure: A literature review," *Landscape and Urban Planning* 81, 3 (2007): 173.

Figure 2.2. Goderich waterfront

Figure 2.3. Demonstration forest, Perth County

down to municipal-level applications as well as in urban, peri-urban, and rural locales of the country. Benedict and McMahon, in *Green Infrastructure: Linking Landscapes and Communities* (2006), describe the linking of natural systems across Florida—the Statewide Ecological Network. The book also describes the natural framework in which planning in Maryland is conducted.

Another publication directed at design for rural areas is *Growing with Green Infrastructure* (Williamson 2003). It lays out the basic design templates for protecting important natural areas in the rural landscape, including areas that protect water supplies, treed landscapes, wetlands, and the connections linking all of these areas together. Connection is important for natural environment protection and enhancement purposes (e.g., biodiversity protection) and as a foundation for a network of off-road trails for human transport. The book also describes the importance of cleaning up contaminated lands— brownfield locations—and designing grey infrastructure (e.g., highways, utility corridors) to consider impacts when bisecting natural grounds. Several examples of plans for areas in Pennsylvania are provided, usually for areas on a watershed-connected basis.

Planning on an environmental foundation has been under way in Europe for a considerable time. Various environmental initiatives of the European Union (EU) require that countries in the union protect their environments within specific timelines and with stated, quantifiable objectives. These various environmental factors are associated primarily with biodiversity protection and climate change protection (European Commission 2010; European Union 2008, 2010, 2012a, 2012b).

The United Kingdom (2012) has recently amended its planning legislation to incorporate the concept of GI. The following planning policy sections highlight the importance of GI in community design:

Policy 99
Local Plans should take account of climate change over the longer term,
including factors such as flood risk, coastal change, water supply and
changes to biodiversity and landscape. New development should be
planned to avoid increased vulnerability to the range of impacts arising
from climate change. When new development is brought forward in
areas which are vulnerable, care should be taken to ensure that risks can
be managed through suitable adaptation measures, including through
the planning of green infrastructure.

Policy 114
Local planning authorities should set out a strategic approach in
their Local Plans, planning positively for the creation, protection,
enhancement and management of networks of biodiversity and green

infrastructure.... Green infrastructure is defined as: A network of multifunctional green space, urban and rural, which is capable of delivering
a wide range of environmental and quality of life benefits for local communities. (ibid., 23, 26)

Several publications from the United Kingdom point to the importance of
GI in community design. These documents have been formulated by Natural
England and the Environment Ministry for the government (Natural England
2009, 2011; United Kingdom and Secretary of State for Environment, Food
and Rural Affairs 2011). Another group, North West England, has also provided a GI guide for community design and development (North West Green
Infrastructure Think Tank 2007).

There are many examples of green infrastructure being used as an important tool in defining a "sense of place" for a community. These examples
include the town of Goderich, in Ontario, Canada, with its central court
square, used as a main community meeting space, place of celebration, and
beautiful public green space (Town of Goderich 2009). Another example of
an important green space defining an adjacent urban area is the Greenbelt
for the Greater Golden Horseshoe, also in Ontario, Canada (Ontario Ministry
of Municipal Affairs and Housing 2005b). The greenbelt serves as an important backbone for the environmental goods and services needed to support
the urban activity of the Toronto-centred region. The greenbelt is also an
important tool for defining the region's distinctive sense of place by utilizing
its agricultural, natural, and rural landscape.

Rural areas can be designed to acknowledge the many attributes defined
in the conceptual framework of green infrastructure presented above. Sandström's (2002) work in Sweden provides substantive documentation of attributes of GI that can support health and wellness attributes and elements of a
community (see Appendix A).

The components of GI have many important goods and services that can
be used to ameliorate the impacts of climate change and peak oil in a rural
community. The following are a sample of the ideas that can assist in building
resilience for rural communities through effective planning.

► Locate development to protect invaluable local water supplies.

► Provide an interconnected natural heritage system that can yield
benefits for nature and serve as a network for active transportation
systems.

► Provide treed landscapes that in turn yield many health and wellness attributes (e.g., air filtration, water absorption and groundwater

retention, anthropocentric noise dampening, shading for summer conditions, etc.).

▶ Provide natural areas and recreational green spaces close to residents to allow equitable access for mental and physical health opportunities.

▶ Provide and promote renewable resource sources for sustainable agricultural, forestry, and water needs.

▶ Provide lands for actual or potential local food production in proximity to rural population areas.

▶ Design landscapes that provide for protection, restoration, and celebration of natural areas as common binding elements for community socialization.

SENIOR GOVERNMENT PLANNING FOR NATURAL SYSTEMS: CASE STUDY WORK IN ONTARIO

The elements of a community planning structure based on natural systems are present in Ontario. Moreover, these planning elements are often associated with GI components that can assist in making rural communities more resilient to anticipated climate change and peak oil impacts.

The senior government in Ontario is the provincial government, which plays a key role in the delivery of legislation, policy direction, and funding with respect to planning. The federal government also becomes involved in planning primarily through funding mechanisms that assist in the implementation of programs. As well, there are some peripheral planning provisions embedded in federal legislation (e.g., the Fisheries Act).

Planning at the rural community level is guided by a host of legislative and policy guide requirements set by the provincial government. Through the federal-provincial constitutional framework, the provincial government is responsible for land-use planning. Municipalities are deemed to be "creatures of the province" under the legislative framework, and Ontario has taken a hands-on approach to planning in them (Hodge and Gordon 2013). This has been especially so with new land-use planning legislative instruments created recently.

The principal tool used by the provincial government to direct land-use planning in Ontario is the Planning Act (Ontario 1990). This act succinctly lays out the "provincial interests" in planning in Section 2, listed as follows:

(a) the protection of ecological systems, including natural areas, features and functions;

(b) the protection of the agricultural resources of the Province;

Figure 2.4. Live snow fence, Waterloo Region

Figure 2.5. Huron County streetscape

Figure 2.6. Sunset at the Pinery, Huron County

(c) the conservation and management of natural resources and the mineral resource base;

(d) the conservation of features of significant architectural, cultural, historical, archaeological or scientific interest;

(e) the supply, efficient use and conservation of energy and water;

(f) the adequate provision and efficient use of communication, transportation, sewage and water services and waste management systems;

(g) the minimization of waste;

(h) the orderly development of safe and healthy communities;

(h.1) the accessibility for persons with disabilities to all facilities, services and matters to which this Act applies;

(i) the adequate provision and distribution of educational, health, social, cultural and recreational facilities;

(j) the adequate provision of a full range of housing, including affordable housing;

(k) the adequate provision of employment opportunities;

(l) the protection of the financial and economic well-being of the Province and its municipalities;

(m) the co-ordination of planning activities of public bodies;

(n) the resolution of planning conflicts involving public and private interests;

(o) the protection of public health and safety;

(p) the appropriate location of growth and development;

(q) the promotion of development that is designed to be sustainable, to support public transit and to be oriented to pedestrians. (6–7)

All of the elements of provincial planning outlined above are impacted by natural systems, and many of these provincial interests are tied to key components of natural systems planning. In considering the conceptual framework of green infrastructure planning, all elements of environmental, social, and economic enterprises in communities are impacted. This can include the productive agricultural lands for employment, the trail networks formed by natural system waterways/shorelines, as well as park spaces that give psychological and physical health benefits to residents.

Another key planning tool provided to municipalities is the Provincial Policy Statement (Ontario Ministry of Municipal Affairs and Housing 2005a). This document forms the overall policy basis on which municipal plans are created across the province. The statement provides direction to planning on a host of matters, and the three key sections relevant to natural systems planning are

Section 1.5—building strong communities through the provision of public spaces, parks and open space

Section 2.1—the wise use and management of natural heritage resources

Section 3.1—the protection of public health and safety through the preclusion of development in natural hazard areas. (ibid., 10, 15, 22)

It is important to note that the Ontario Planning Act requires that all municipal decision making on land-use planning *be consistent* with the policies of the Provincial Policy Statement.

In addition to general policy documents that apply across the province, the Ontario government has enacted specific planning legislation for areas deemed to be of special land-use significance. These special geographic landscapes have a natural features orientation, and it is interesting to note their existence and the policies that protect them, such as the Niagara Escarpment Planning and Development Act (Ontario 1973) and the Oak Ridges Moraine Conservation Act (Ontario 2001). All of these acts have provincial plans associated with them, and all have a natural systems protection and enhancement theme.

In addition to specific area plans, the provincial government has numerous legislative pieces and policy documents that aim to protect natural systems. Municipalities, when creating resilient natural planning frameworks for their jurisdictions, are required to look at these plans as important guideposts for the future.

This web of documents and policies illustrates the movement toward a "biophilic mindset" in the protection of our life-giving natural systems while also using Ontario's rich resource base for economic, psychological, and social needs. The following elements of natural systems planning at the provincial level are listed in their historical sequence, which stretches over the past few years:

▶ Greenbelt Act and Plan (Ontario 2005; Ontario Ministry of Municipal Affairs and Housing 2005b)

▶ Clean Water Act and the associated Source Water Protection Program and plans for conservation authority watersheds across the province (Ontario 2006)

▶ Species at Risk Act (Ontario 2007)

▶ 50 Million Trees Program (Ontario Ministry of Natural Resources 2007)

▶ Lake Simcoe Protection Act and Its Associated Plan (Ontario 2008)

▶ Water Opportunities and Water Conservation Act (Ontario 2010)

▶ Climate change action planning through the documents *Go Green: Ontario's Action Plan on Climate Change* and *Climate Ready: Ontario's Adaptation Strategy and Action Plan 2011–2014* (Ontario Ministry of the Environment 2011a, 2011b)

▶ Biodiversity protection strategy as outlined in the publication *Biodiversity: It's in Our Nature* (Ontario Ministry of Natural Resources 2012)

As can be seen from the variety of planning interests and the rapid succession of new legislation in recent times, a greening perspective to planning is in vogue in the senior level of government in Ontario. This sets the foundational elements for planning at the local level that can comprise a central GI planning structure for resilient community development.

Although there are other pieces of government legislation and policy direction (e.g., Green Energy and Economy Act), the above documents are illustrative of a greening perspective that can and should be applied to land-use planning. Many of the action items included in these documents can assist in building resilience to climate change and peak oil impacts in rural communities.

▶ A clean water protection planning framework has a local orientation for putting operational and land-use design controls in place to protect "local" drinking water resources (i.e., to work within the carrying capacity of fresh water in a local watershed area). Water conservation legislation is also intended to protect the existing supply of clean water for Ontarians.

▶ The protection of "species at risk" habitat promotes the retention of endangered and at-risk species, which works toward a healthy and diverse ecosystem for Ontario.

▶ The 50 Million Trees Program is designed to assist the United Nations Billion Tree world initiative to fight climate change.

▶ The Lake Simcoe plan is concerned with the assimilative capacity of the lake to take additional nutrients from the surrounding land base and to put measures in place to control nutrient loading into the water basin (i.e., control development and rural land use activity causing pollution in the local watershed area).

▶ The climate change action initiatives are directed at reducing the carbon footprint of development in Ontario and are directly associated with the climate change mitigation and adaptation agenda.

▶ The biodiversity protection strategy for the province is a big picture need and calls for action to protect the biosphere. This document portrays the key interrelationships between humans and the natural environment and lays out the various goods and services (provisioning, regulating, cultural, supporting) of nature. Important mechanisms for building resilience to climate change and peak oil impacts can be discerned from the text, such as provision of essentials of life (clean soil, water, air), potential for storm event buffering and protection, areas set aside because of natural hazard conditions, and linked natural systems useful for active transportation systems.

In addition to legislative and policy derivation mechanisms in the provincial "top down" planning system, the Canadian federal government also from time to time offers funding incentives to implement public interest initiatives. In the environmental area, there are many instances in which third-party groups are used to assist in implementation on the ground (e.g., an environmental farm plan). This is illustrated in the next chapter dealing with local municipal case studies.

CONCLUDING THOUGHTS

Using nature as a central design premise for resilient communities is a relatively new concept. It builds on the narrower planning practices of the past to plan for some open-space areas within a community (for park and recreational space) and not to develop on lands that would be subject to catastrophic events such as erosion, flooding, or slumping. The notion of using the goods and services embedded in nature that are freely given to us humans is a newer concept that reflects a greater appreciation and acknowledgement of nature that can assist us in our planning work today.

Senior levels of government have been moving slowly toward acknowledging the importance of nature and natural systems as key design tools for building resilient communities. In examining the evolution of planning legislation in Ontario over the past twenty years, we can see a greater acknowledgement of the importance of nature and natural elements in planning. It is reasonable to assert that, as we move forward in attempting to ameliorate the challenges of climate change and peak oil, the use of nature and natural systems will become even more embedded in the ways of doing things.

The next chapter outlines mechanisms of nature that municipalities and local organizations use to build resilience into their local planning and development efforts.

Notes

1   This philosophical approach to planning is discussed in detail in the Environmental Commissioner of Ontario's 2006–07 annual report to the legislature, *Irreconcilable Priorities: The Challenge of Creating Sustainable Communities in Southern Ontario*.

References

Benedict, M.A., and E. McMahon. 2006. *Green Infrastructure: Linking Landscapes and Communities*. Washington, DC: Island Press.

Caldwell, W. 2011. *Rediscovering Thomas Adams*. Vancouver: UBC Press.

Carrick, P. 2012. "Aldo Leopold's Concept of Land Health: Implications for Sound Public Health Policy." In *Human Health and Ecological Integrity: Ethics, Law, and Human Rights*, edited by L. Westra, C.L. Soskolne, and D.W. Spady, 56–66. London: Routledge.

Center for Neighborhood Technology. 2010. *The Value of Green Infrastructure: A Guide to Recognizing Its Economic, Environmental, and Social Benefits*. http://www.cnt.org/ repository/gi-values-guide.pdf.

Daniels, T., and K. Daniels. 2003. *The Environmental Planning Handbook for Sustainable Communities and Regions*. Chicago: American Planning Association.

Environmental Commissioner of Ontario (ECO). 2007. *Irreconcilable Priorities: The Challenge of Creating Sustainable Communities in Southern Ontario*. ECO Annual Report, 2006–07. Toronto: ECO.

European Commission. 2010. *Green Infrastructure*. Luxembourg: European Commission Publications Office.

European Union. 2008. *Building Green Infrastructure for Europe*. Brussels: European Environmental Bureau.

——. 2010. *Building Up Europe's Green Infrastructure: Addressing Connectivity and Enhancing Ecosystem Functions*. Brussels: European Union.

——. 2012a. "Environment: Green Infrastructure." http://ec.europa.eu/environment/nature/ ecosystems/index_en.htm.

——. 2012b. "European Union." http://europa.eu/index_en.htm.

Green Infrastructure Ontario Coalition. 2012. *Health, Prosperity, and Sustainability: The Case for Green Infrastructure in Ontario*. http://www.greeninfrastructureontario.org/ report.

Hodge, G., and D.L.A. Gordon. 2014. *Planning Canadian Communities: An Introduction to the Principles, Practice, and Participants*. 6th ed. Toronto: Nelson.

Leopold, A. 1949. *A Sand County Almanac: With Other Essays on Conservation from Round River*. London: Oxford University Press.

McHarg, I. 1969. *Design with Nature*. Garden City, NY: Doubleday and Natural History Press.

Natural England. 2009. *Green Infrastructure Guidance.* http://publications.naturalengland.org.uk/publication/35033.

———. 2011. *Think BIG: How and Why Landscape-Scale Conservation Benefits Wildlife, People, and the Wider Economy.* http://publications.naturalengland.org.uk/publication/30047.

North West Green Infrastructure Think Tank. 2007. *North West Green Infrastructure Guide.* Manchester, UK: North West Green Infrastructure Think Tank.

Ontario. 1973. *The Niagara Escarpment Planning and Development Act.* Toronto: Queen's Printer.

———. 1990. *The Planning Act,* RSO 1990. Toronto: Queen's Printer.

———. 2001. *The Oak Ridges Moraine Conservation Act.* Toronto: Queen's Printer.

———. 2005. *The Greenbelt Act.* Toronto: Queen's Printer.

———. 2006. *The Clean Water Act.* Toronto: Queen's Printer.

———. 2007. *Endangered Species Act.* Toronto: Queen's Printer.

———. 2008. *Lake Simcoe Protection Act and Its Associated Plan.* Toronto: Queen's Printer.

———. 2010. *Water Opportunities and Water Conservation Act.* Toronto: Queen's Printer.

Ontario Ministry of the Environment. 2009. *Lake Simcoe Protection Plan.* Toronto: Queen's Printer.

———. 2011a. *Go Green: Ontario's Action Plan on Climate Change.*

———. 2011b. *Climate Ready: Ontario's Adaptation Strategy and Action Plan 2011–2014.*

Ontario Ministry of Municipal Affairs and Housing. 2005a. *Provincial Policy Statement.*

———. 2005b. *Greenbelt Plan.*

Ontario Ministry of Natural Resources. 2007. *50 Million Trees Program.*

———. 2012. *Biodiversity: It's in Our Nature.*

Randolph, J. 2012. *Environmental Land Use Planning and Management.* Washington, DC: Island Press.

Sandström, U.G. 2002. "Green Infrastructure Planning in Urban Sweden." *Planning Practice and Research* 17, 4: 373–385.

Rouse, David C., and Bunster-Ossa, Ignacio F. 2013. *Green Infrastructure: A Landscape Approach.* American Planning Association, Planning Advisory Service Report Number 571. Chicago, IL.

Town of Goderich. 2009. *Official Plan.* http://www.goderich.ca/en/townhall/Planning.asp.

Tzoulas, K., K. Korpela, S. Venn, V. Yli-Pelkonen, A. Kaźmierczak, J. Niemela, and P. James. 2007. "Promoting Ecosystem and Human Health in Urban Areas Using Green Infrastructure: A Literature Review." *Landscape and Urban Planning* 81, 3: 167–78.

United Kingdom. 2012. *National Planning Policy Framework.* https://www.gov.uk/government/uploads/system/uploads/attachment_data/file/6077/2116950.pdf.

United Kingdom and Secretary of State for Environment, Food and Rural Affairs. 2011. *White Paper on the Natural Choice: Securing the Value of Nature.* https://www.gov.uk/government/uploads/system/uploads/attachment_data/file/228842/8082.pdf.

Williamson, K. 2003. *Growing with Green Infrastructure.* Doylestown, PA: Heritage Conservancy.

Appendix A: Sandström Criteria for Green Infrastructure Goods and Services Planning (from Sandström 2002, 43–57, with slight wording modifications to reflect a rural context rather than an urban one)

Recreational Criteria

Importance to everyday life—daily use by citizens for walking, exercising, playing, and social interaction

Accessibility—location of green spaces within walking distance and without barriers (e.g., roads with heavy traffic)

Geographical distribution—fair distribution of green spaces in all areas of the community

Interconnectedness—availability of greenways between green spaces

Pedagogy—availability of green spaces for school excursions and providing understanding of nature

Public health—improves quality of life and promotes healthy habits

Surface water—presence of lakes, ponds, and streams improves the quality of green spaces

Appreciation—different ways in which people appreciate parks, woods, and other green spaces

Number and size of green spaces—parks and other green spaces

Aesthetic functions—role of parks and other green spaces to beautify the community

Public-private green spaces—private gardens as an important complement to public green spaces

National interest—preservation of specific green spaces of importance for national heritage

Allotments—leasehold of small plots to grow flowers and vegetables

Biodiversity Criteria

Biodiversity (ecosystem level)—multiplicity of ecosystems in the rural environment

Biodiversity (species level)—presence of a great variety of native species in the rural environment

Biodiversity (landscape level)—variation of landscapes in the overall visual landscape

Presence of greenways—green passageways between habitats, including connection to the community's land to facilitate migration of species

Valuable green cores—green spaces with native habitats that can act as breeding grounds for species

Importance of surface water—bodies of surface water increase ecosystem and species diversity

Green space management—a green plan has clearly stated management criteria for promoting biodiversity

Size of green spaces—positive correlation between the size and number of green spaces and species

Habitat continuity—older habitats develop higher species diversity compared with younger ones

Rare/threatened habitats—importance of preserving rare/endangered species habitats and species

Barrier effects—human-made obstructions in the landscape that prevent migration of species between habitats

Scientific values—habitats of specific scientific values

Fragmentation and edge—subdividing a continuous habitat affects smaller entities, increases the number of ecotones and species, and has impacts on local climate

Representativity—habitats representative of a particular landscape

Metapopulation aspects—aggregates of patch populations in the community landscape

Rural Community Structure Criteria

Identity and character—each community has characteristic green spaces that citizens recognize as important and unique

Structuring functions—lines of trees, avenues, and other vegetation along streets, roads, and squares

Discerning component—a community becomes more comprehensible for citizens because green spaces separate various landscapes in the community into smaller districts

Unifying factor—green spaces unite the community in a natural way

Linkage to surrounding hinterland—green spaces provide natural links between the community and surrounding landscape

Cultural Identity Criteria

Specific cultural aspects—single cultural features, such as cemeteries or mansion parks, not included in other indicators

Historical heritage—green spaces of historical importance, associated with special historical events

District features—preserving ecosystems developed especially in particular landscapes of the community

Community character—historical green planning is mirrored in the existing community landscape

National cultural interests—green spaces of national cultural value

Local traditions—green spaces that result from particular cultivation traditions and/or techniques

Environmental Factor Criteria

Filter pollutants—deciduous and other trees can act as filters and clean the air

Protection zones—vegetation shields houses and squares from wind

Improve local climate—vegetation increases humidity, cools down the landscape, and provides shaded areas

Ventilation system—by leading fresh air from the open space surroundings, greenways exchange and thereby improve air quality in the community

Noise reduction—vegetation reduces the effects of noise

Biological Solutions to Technical Problems Criteria

Cleaning storm water—green spaces can prevent polluted rainwater from running directly into a recipient body of water

Recipient for organic waste—possibilities to take care of organic waste in green spaces

Importance for sustainability—green spaces as important elements in local sustainable development policies

CHAPTER THREE

PLANNING WITH NATURE: EXPERIENCES AT THE GRASSROOTS LEVEL

PAUL KRAEHLING AND WAYNE CALDWELL

This chapter builds on the discussion in Chapter 2 of the use of nature and natural systems to build resiliency to face the challenges of climate change and increasing energy prices (peak oil) in rural areas. We present case studies of rural municipalities in southern Ontario that use elements of nature and natural systems—the "green infrastructure" of communities—to address current and anticipated challenges (Caldwell 2013a).

Green infrastructure (GI), as discussed in Chapter 2, is defined as

natural vegetation and green technologies that collectively provide society with a broad array of products and services for healthy living. Green infrastructure takes many forms including but not limited to the following: ... forests, natural areas, greenways, streams/water bodies and riparian zones, meadows and agricultural lands; green roofs and green walls; parks, gardens and landscaped areas, community gardens, and other green open spaces; rain gardens, bioswales, engineered wetlands and storm water ponds.

Green infrastructure also includes soil—in volumes and qualities adequate to sustain living green infrastructure and absorb water, as well as technologies like porous paving, rain barrels, cisterns and structural soil.

(Green Infrastructure Ontario Coalition 2012, 2; with slight word modifications)

The identified features of GI, including economic, social, and environmental features, provide many elements for human health and well-being. For example, they provide the protection of surface and groundwater quality, the provision of water supply, groundwater recharge, increased health opportunities through outdoor recreation, improved air quality, reduction in summer heat island effects, reduced flooding, local agriculture production (urban and rural), improved community cohesiveness, reduction in traditional "grey infrastructure" servicing needs (storm sewers, impervious surfaces), opportunities for community education, community amenity and attractiveness, reduced energy needs, improved wildlife habitat, reduced noise pollution, provision of resources (where available) such as lumber and aggregates, and climate change mitigation (Center for Neighborhood Technology 2010).

Several recent research undertakings by Dr. Wayne Caldwell and graduate students in the University of Guelph's School of Environmental Design and Rural Development have examined aspects of using nature/natural systems in furthering rural community sustainability in Ontario (Caldwell 2008, 2010, 2011, 2013a, 2013b). This chapter captures several of these examples as we examine the opportunities to plan and build resilient rural communities through a rural land-use lens.

When considering the use of green infrastructure for municipal planning, we need to keep in mind the notion of direct payback to the community's budget— either through cost savings in operating expenditures (now or in the future) or through a better return on investment in the capital outlay for new infrastructure. In addition, new community job opportunities are often associated with the provision of green infrastructure. These notions relate to the general anthropocentric betterment orientation of the community planning function.

The case studies that follow are illustrative of nature and natural systems planning in southern Ontario rural municipalities. It is hoped that, if several such actions are stitched together in an overall comprehensive plan for a municipality, then there is the prospect that resiliency to future challenges to a community (climate change, peak oil conditions) can be better addressed.

LOCAL RURAL GOVERNMENT AND LOCAL COMMUNITY-LEVEL CASE STUDY WORK

The following discussion highlights efforts that municipalities and local organizations are using in southern Ontario to use nature and natural systems to plan more resilient communities. The GI elements in many instances address directly or indirectly mitigation of and/or adaptation to climate change as well as impacts of increased energy prices (peak oil). The examples are illustrative of the robust variety of nature-based systems used to promote environmental protection and enhancement in rural communities, and we discuss them under the following general topic areas:

(1) general community capacity development approaches to furthering environmental initiatives and focusing attention on the healthy environment agenda, as shown by the Town of Caledon municipal environmental orientation and Carolinian Canada's the Big Picture, an interconnected natural heritage system plan;

(2) stewardship guides and programs for farmers and non-farmers where environmental stewardship has been shown to be beneficial

to all, focusing on the Environmental Farm Plan and associated government programs and the *Ontario Rural Landowners Stewardship Guide;*

(3) on-the-ground tree-planting success stories, exemplified by Trees for Mapleton, Munsee-Delaware First Nation Tree Planting and Sustained Harvesting Initiative, and the Green Legacy initiative by the County of Wellington; and

(4) acknowledging the goods and services of nature through payment to provisioning landowners, as highlighted in the Alternative Land Use Systems (ALUS) program, and tying into senior government financial incentive programs such as municipal tax class provisions for protecting significant natural heritage lands and Species at Risk land and water habitat protection incentives.

ENVIRONMENTAL PLANNING AT THE LOCAL COMMUNITY LEVEL

The Town of Caledon – Caledon, Ontario, is a municipal leader in environmental planning. It has taken a green infrastructure approach to planning whereby all proposed actions in the community are put through an environmental impact lens.

The environmental focus of the Town of Caledon is evidenced in its Official Plan, with a focus on environmental sustainability, and council has enacted an Environmental Advisory Committee of interested citizens to give councillors advice on all things dealing with the environment (Caldwell 2008). The town has also created an environmental planner position, rare for a municipality with under 100,000 people (ibid.).

Carolinian Canada: The Big Picture – Carolinian Canada, a non-profit environmental organization, has identified the need to create an interconnected network of green spaces across southern Ontario, referred to as the Big Picture (Carolinian Canada 2000). This organization has thought through the need to have this green infrastructural framework serve as a backbone for natural area protection across the Carolinian eco-zone of Ontario. The creation of this comprehensive natural heritage framework depends on many partners—municipalities, private landowners, non-profit organizations, and businesses—to assist in implementation.

General Observations – These examples of organizations attempting to build more resilient natural systems are illustrative of attempts to be conscious of climate change and peak oil impacts on the ground at the local level.

ENVIRONMENTAL STEWARDSHIP AT THE INDIVIDUAL PROPERTY LEVEL

This section highlights the two guides that have been used for empowering individuals to make environmental improvements and increase economic sustenance on personal property.

The Environmental Farm Plan and the
Canada-Ontario Farm Stewardship Program

As succinctly stated on the Ontario Ministry of Agriculture, Food and Rural Affairs website, "Environmental Farm Plans (EFP) are assessments voluntarily prepared by farm families to increase their environmental awareness in up to 23 different areas on their farm. Through the EFP local workshop process, farmers will highlight their farm's environmental strengths, identify areas of environmental concern, and set realistic action plans with time tables to improve environmental conditions. Environmental cost-share programs are available to assist in implementing projects" (OMAFRA 2013). In terms of the environmental initiatives that can be completed under the EFP, the following is a sampling: best management practices for storm water management, water body riparian zone protection, environmentally sensitive area protection for upland pastures, groundwater protection, shelterbelt and native vegetation establishment, irrigation management and erosion control, cover crop planting, and invasive plant protection.

Since 1993, 35,000 farms have gone through the program to encourage good environmental practices that respect farming and improve farm viability in Ontario (OMAFRA 2013). The program consists of a workshop, a self-directed workbook, and action priority sheets peer-reviewed by trained experts in the field. It is a voluntary "stewardship" initiative with partial funding from senior government levels to motivate activity. In 2013, a new round of senior-level government funding was approved through the Growing Forward program. With this program, funding of up to $30,000 was made available to each participating farmstead, resulting in significant environmental enhancement as well as increasing other farm viability initiatives (Ontario Soil and Crop Improvement Association 2013).

In terms of measuring the success of the program, Prairie Research Associates (2011) were contracted by senior government levels to measure outcomes from a recent round of the Environmental Farm Plan program in 2010. The survey showed that the majority of farmer efforts resulted in improvements to soil quality, water quality, and family health and safety. On average, $54,000 was spent to make on-farm improvements, with government funding representing approximately a third of the total amount.

This program is illustrative of a partnership format between the government and private landowners to make environmental enhancements a priority.

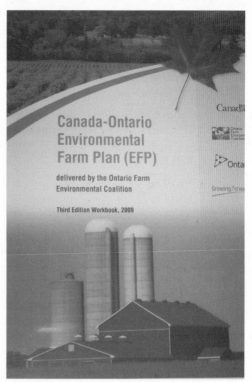

igure 3.1. Caledon Green Fund program   Figure 3.2. Ontario Environmental Farm Plan

igure 3.3. Ontario Rural Landowner Stewardship Guide

The Ontario Rural Landowners Stewardship Guide – This guide is an environmental stewardship "best practices" manual developed from a similar 2006 guide for water quality protection issues in the Ausable Bayfield Conservation Authority area along Lake Huron. The guide, patterned after the Environmental Farm Plan, has fourteen sections with over 100 action items to improve environmental conditions for rural non-farm properties.

In 2012, a major research effort assessed the value of the guide for rural landowners in initiating environmental stewardship actions (Caldwell 2013b). The research indicated that the guide had morphed into at least ten other specific area guides for parts of Ontario and that it was being used by several conservation authorities and stewardship organizations (e.g., Carolinian Canada, Equine Guelph). In terms of quantifying the use of the guide, it was determined that approximately 100 workshops had been held by stewardship organizations since 2007 and that over 2,000 participants had used the guide through these workshop experiences (ibid.).

In an attempt to quantify on-the-ground environmental actions that had taken place, the experiences of the Lake Simcoe Community Stewardship Program were noted. Under the auspices of the Lake Simcoe Water Protection Plan 2009, funding from senior-level government agencies was provided for stewardship initiatives of $1 million annually for a three-year period. This funding was matched by other agencies and private landowners to return a total value of environmental initiatives to approximately $10.4 million. Many of the environmental enhancements involved shoreline protection work and septic system updating (Dufferin-South Simcoe Lake Stewardship Network 2011).

General Observations – It has been illustrated through the use of the rural area stewardship guides—one for farmers and one for non-farmers—that incentives directly affect implementation. In addition, while providing payback to the participant either in immediate operational cost savings or in longer-term property value enhancements, there are environmental improvement dividends paid. In the guides are many suggestions for making changes to a property to be climate-change adaptive and cognizant of energy use (to reduce the need for fossil fuels) (Caldwell 2013; OMAFRA 2013). The reader is encouraged to examine the *Ontario Rural Landowners Stewardship Guide* at www.stewardshipmanual.ca.

TREE-PLANTING INITIATIVES

Trees for Mapleton – This case comprises a grassroots stewardship council working with local governments and non-profit agencies to participate in tree-planting activities in the Township of Mapleton. This township is located within Wellington County, a prime agricultural growing area of south-central Ontario. The majority of the landscape comprises fields for agricultural production. To assist in improving crop production returns, reducing rising energy costs, and improving local biodiversity, a concerted tree-planting effort has been under way for the past decade (Ferguson 2012).

To undertake work, farmers prepare environmental farm plans to document the priorities for on-farm tree-planting initiatives. In many instances, economic benefits are illustrated by the strategic planting of trees to serve as windbreaks for fields, shelterbelts to protect farm buildings, and living snow fences along roadways.

· The tree plantings are adding to an interconnected tree network across the township. To date (since 2006), 400,000 trees have been planted, and over 300 kilometres of windbreaks have been provided (Caldwell 2013a).

The planting of trees has been identified as a useful green infrastructural approach to promoting resiliency in the countryside and addressing peak oil and climate change challenges. The windbreaks have multifunctional benefits to humans and nature, including improving soil conditions (moisture, organic materials, topsoil, nutrients); providing ecosystem variety to monoculture crops by bringing in natural insect pest predators (birds) to the area; increasing economic returns by increasing crop yields in areas adjacent to the windbreaks (because of earlier germination of seeds and protection from weather during the crop-growing season); and serving as natural snow fences that reduce snow management costs on municipal roadways (County of Wellington 2013).

The planting of trees also provides benefits to nature and humans. It increases the connectivity of natural systems, thereby promoting biodiversity transference between areas. It increases soil health and reduces soil erosion, thereby reducing the need for fuel-based fertilizer supplements for crop production. The program serves as a community capacity development strategy uniting people around the cause of improved environmental conditions.

The Munsee-Delaware First Nation Reforestation Project – This case study illustrates full-cycle sustainability of tree planting with environmental, economic, and social benefits. The reforestation project occurred on idle agricultural lands on the Munsee-Delaware First Nation's land near London, Ontario, in 2010. The project involves several interrelated aspects associated with nature as an environmental benefit, an economic generator, and a social capacity development mechanism (Caldwell 2011).

As an environmental benefit, the project saw the planting of over 70,000 trees on a vacant piece of land of forty-eight hectares. The fast-growing, hybrid poplar trees provide significant ecological benefits, including oxygen production, water retention, erosion prevention, local climate moderation, and carbon sequestration.

Figure 3.4. Trees for Mapleton

Figure 3.5. Treeplanting in Mapleton Township

From an economic perspective, the greatest benefit comes in two forms. The trees have been planted with a sustainable biofuel purpose, so after thirty-one years (near the end of their life span) they will be harvested for energy production purposes at a nearby electrical generation facility (ibid.). Over the course of the trees' life span, a major Canadian bank has agreed to pay the First Nation band an annual amount for the carbon sequestration impacts, and this money is provided to meet the bank's objective of being a carbon-neutral enterprise.

From a social development perspective, the trees are an investment in providing new maintenance employment opportunities for younger people in the First Nation band. There is also community pride in turning unused land into a valuable asset to be shared by the community.

This case is interesting in that it is believed to be the first carbon forest to be developed on First Nations land in Ontario and the first time that a comprehensive guide has been used that accounts for the greenhouse gas sequestration value of the trees (ibid.). It illustrates not only how the impacts of climate change can be mitigated by one project at a time but also a mechanism for future fuel production that is not fossil fuel–based. The First Nation's intent is to replicate this project on other lands in the province and to view tree reforestation with biofuel harvesting as a continuing project over multiple generations.

The Green Legacy Program – This is a "best case" example of a local municipality spurring on the planting of landscaping materials for sustainability purposes. The program comes out of Wellington County in south-central Ontario and involves an annual tree/shrub-planting program. The program evolved from the county's 150th anniversary event in 2004 to a program now recognized as the largest municipal tree-planting program in North America (County of Wellington 2013).

During the first year, 150,000 trees were planted in various locales across the county. By 2013, over 1.3 million trees and shrubs had been planted, two municipally owned tree nurseries have been established, and the efforts of the county have been recognized provincially, nationally, and internationally (ibid.). It is noteworthy that the United Nations has recognized the program in its efforts to fight climate change.

Germinating seeds, tending to seedlings, and planting of trees and shrubs are completed at modest cost to the county administration, with community volunteers assisting in all aspects of these endeavours. Schoolchildren, through environmental education efforts with local school boards, are involved in all aspects of propagating and planting the landscaping materials. Property owners and organized volunteer groups arrange to plant the trees and shrubs in areas across the county. The Trees for Mapleton program is a component of the Green Legacy program.

As a result of the Green Legacy program, plantings have been made to beautify roadway corridors and private/public properties and to provide significant ecological goods and services for nature and humans (e.g., riparian buffers to streams, landscape corridor networks, shade and air quality benefits, biodiversity enhancement, etc.). The benefits of tree planting to human and environmental health have been communicated to the public through education efforts along the way.

GOODS AND SERVICES PAYMENTS FOR ENVIRONMENTAL PROTECTION OR ENHANCEMENT MECHANISMS

This discussion illustrates instances in which the goods and services of a specific natural element of an environmental initiative are compensated either through direct payments from some source or through government tax system considerations.

To understand the necessity of making payments from a public "funding source" to the provision of goods and services that nature otherwise provides for free, the reader is referred to a recent work by the David Suzuki Foundation, which presents research on the value of the greenbelt circling the Toronto-centred region (Greenbelt Foundation 2008). This report puts the value of nature in the greenbelt at $2.6 billion annually in non-market economic value.[1] The study raises an interesting question about how our economic system can "afford" payment for services of such extraordinary valuation.

The Alternative Land Use Services Program in Norfolk County – The Norfolk Land Stewardship Council spearheaded the formulation of this program in conjunction with the Norfolk Federation of Agriculture in Norfolk County (located on the northeast shore of Lake Erie) (ALUS 2013). ALUS, the Alternative Land Use Services program, is a means by which farmers are assisted in conserving and restoring natural ecological capital on their farms. The program is incentive based to pay farmers who set aside land for natural features such as grasslands, forests, wetlands, and habitats for species at risk and who protect their continued functioning. The program was funded on the beliefs that individual farmers should not have to bear the costs of being good environmental stewards of the land and that, because there is a broader public interest in environmental protection and enhancement, some public financial incentives should be made available (Agriculture and Agri-Food Canada and North American Wetlands Conservation Council–Canada 2009). A pilot project was funded from 2008 to 2011 by various non-profit groups, various government levels, businesses, and farmers in the area. In a nutshell, the program provided small annual payments (i.e., $150 per acre) for land that has been improved for natural purposes within a set of specific conditions. In the first two years of the program, 450 acres of land were improved for natural purposes (Caldwell 2010).

This program is reflective of a grassroots, rural, stakeholder-based enterprise that provides "a fee for natural service." The program is interesting since it does portray mechanisms to address climate change and peak oil impacts. The program has a local community foundation, and it is providing additional natural capital resiliency through increased biodiversity in the area.

Payment by the Government for Goods and Services of Nature – There are a number of instances in which senior levels of government provide financial incentives for the protection of natural systems. These incentives can be classified in three broad categories: (1) income tax breaks for the gifting or easement of lands of ecological sensitivity; (2) property tax breaks for lands used for environmental conservation purposes; and (3) incentives paid as part of different government programs that come and go over time.

The provision of incentives is tied to a variety of conditions that cannot be elaborated here because of the complex details involved. Suffice it to say that the valuation of incentives is modest in terms of the value that the goods and services of nature on private property return to general society.

Among various pieces of legislation in Ontario over the past several years is a general pattern of financial incentives and tax breaks provided when new legislation is introduced. For example, with the introduction of the Endangered Species Act in 2007, a whole series of incentives was provided to encourage landowner interest and participation in protecting habitat. These incentives included the Species at Risk Stewardship Fund, the Species at Risk Farm Incentive program, the Conservation Land Tax Incentive program, and the Managed Forest Tax Incentive program. It is speculated that, because of general economic difficulties at present, the provincial government has had to make cutbacks to funding for these programs.

It is uncertain whether there will be future funding mechanisms from senior levels of government to pay for nature goods and services that ameliorate impacts of climate change and peak oil conditions. As this discussion has shown, limited steps have been made to fund environmental initiatives.

CONCLUDING THOUGHTS

Rural communities are often faced with significant challenges—from stagnant or declining assessment bases to public infrastructure that is often in need of replacement or repair. It is becoming increasingly apparent that new tools and adaptations are necessary to deal with these existing challenges and anticipated new ones (e.g., infrastructure to withstand the increased intensity of storm events brought about by climate change). As this chapter has illustrated, one effective tool is green infrastructure—using the goods and services embedded in nature and natural systems to plan and build healthy and resilient rural communities.

The advantages of using nature and natural systems in our community resiliency planning are multifold: the services provided by nature are free in terms of economic system accounting; they address both short-term and long-term health and wellness propositions for the community; and they often have associated payoffs in terms of social benefits.

The twin challenges of climate change and peak oil can be considered "wicked problems" for future planning (Rittel and Webber 1973). These types of problems involve issues of deriving common understanding and urgency, assessing the need or ability to solve them, determining who is responsible to deal with them, dealing with irreconcilable differences of opinion among stakeholders, and, should the problems be addressed, solving other problems created. It is within this issue maelstrom that the notion of using nature and natural systems in rural community contexts can assist in mediating impacts and aiding planning for adaptive mechanisms to address the issues.

As identified in this chapter, nature and natural systems can be used in a variety of ways to build adaptive and resilient capacity for rural community health and well-being. In many instances, this capacity can be achieved by collaboration and systems thinking long term. The following points illustrate some of the ways in which it can be achieved.

▶ Maintain or enhance tree cover. The many goods and services embedded in trees that often go unnoticed but are essential to our life and well-being include carbon sequestration, water table maintenance with storm water management, oxygen production, shading provision, psychological well-being, erosion prevention, and wind reduction.

▶ Provide ecosystem networks to support biodiversity retention and possible enhancement. Riparian zones along natural waterways allow plants and animals to traverse the landscape and thereby enhance their resiliency to changes taking place. These networks are also useful for alternative transportation corridors for human transport, set the basic structure for community "sense of place" elements, and provide buffers to extreme storm events.

▶ Areas prone to flooding or unstable soil conditions can be set aside from development and used for natural green space "goods and services" activities. Additional buffers will be required along waterways and shorelines for anticipated floods in the future.

▶ Utilize green space as a means to create jobs for local residents (e.g., crop harvesting on an annual basis [food for local or export consumption] or a multiple-decade basis [biofuel tree production]).

▶ Protecting and enhancing natural areas can assist in providing long-term resource availability, if required. The resources in wildlife, fisheries, and forested areas can be used on a sustainable-harvest basis to provide for local needs. In addition, with good attributes, natural assets can serve as employment opportunities in tourism.

▶ Utilize the positive values of nature as a community capacity-building enterprise. Civic tree-planting programs for beautification are a recognized "best practice" to bring people together and leverage action between community and private interests. There are many charitable foundations and granting organizations engaged in civic beautification and nature enhancement.

▶ Use green infrastructure to clean up polluted former industrial sites (brownfields) for new activities or to beautify other "eyesores." Funding programs are available at the senior government levels and from the Federation of Canadian Municipalities (Green Fund Initiatives) to assist in this regard.

▶ Areas set aside for non-development can be recognized as and reused for biodiversity enhancement and natural goods and services protection (e.g., water conservation, species at risk protection, etc.).

▶ Green open spaces used for passive recreation (hiking, viewing nature, picnicking) and active recreation (sporting events) are essential for the health and well-being of community residents. The strategic placement of these facilities in all areas of the community is important for equity of access for all inhabitants.

▶ Recognize the mental and emotional well-being effects of nature and plan accordingly for the provision of open spaces for community residents. Recognize that this resource base is especially important for those in stressful situations or where rapid change is taking place in people's lives.

▶ Open spaces provided through community land-use plans can be used to grow food that in turn can be sold in local community facilities.

▶ Municipalities can delineate through community planning open countryside spaces that should be protected for their cultural heritage and scenic values. In this way, the distinctive "sense of place" of the community through its distinctive landscape can be recognized and protected.

Our perspective is that in working together at all levels—governments, residents, businesses—rural communities can assist in addressing the challenges before us from climate change and peak oil. The protection of existing natural resources, the provision of new green infrastructure in the community, the provision of funding incentives from senior levels of government to kick-start action, and community planning efforts around the topic of nature will all assist in building capacity in the community to withstand the challenges that lie ahead.

Notes

1   The study (Greenbelt Foundation 2008) regarding the goods and services of nature within its area quantifies, in a "preliminary and course-scale natural capital account," the following functions: air quality, climate regulation via stored carbon and annual carbon uptake, flood control, water regulation through runoff control, water filtration, erosion control and sediment retention, soil formation, nutrient cycling, water treatment, plant and agricultural crop pollination, natural regeneration, biological control through pest management, habitat/refugia, recreation and aesthetics, and cultural/spiritual. The report points out that the valuation has been prepared "to encourage others to consider the value of natural capital and its ecosystem services, as well as to stimulate a growing dialogue regarding the real value of natural capital, ecosystem services, stewardship and conservation."

References

Agriculture and Agri-Food Canada and North American Wetlands Conservation Council (Canada). 2009. "Exploration of the Ecological Goods and Services Concepts and Options for Agri-Environmental Policy." Proceedings of meeting in Ottawa, 29–30 April.

ALUS. 2013. "Alternative Land Use Services." http://www.norfolkalus.com/.

Caldwell, W. 2008. Sustainable Rural Communities—Environmental Planning and Innovation: Best Practices for Rural Communities. Guelph: University of Guelph.

——. 2010. Resource Materials for Community Economic Development: Prospering with a Stable or Declining Population. Guelph: University of Guelph.

——. 2011. Mechanisms to Build Resiliency and Mitigate Impacts to Climate Change and Peak Oil while Creating Jobs for Communities in Midwestern Ontario. http://www.workgreen. ca/content/climate-changepeak-oil-views.

——. 2013a. Peak Oil and Climate Change: A Rural Community Guide. Guelph: University of Guelph.

——. 2013b. Ontario Rural Landowners Stewardship Guide. http://www.stewardshipmanual. ca.

Carolinian Canada. 2000. The Big Picture Project. http://caroliniancanada.ca/legacy/ ConservationPrograms_BigPicture.htm.

Center for Neighborhood Technology. 2010. The Value of Green Infrastructure: A Guide to Recognizing Its Economic, Environmental, and Social Benefits. http://www.cnt.org/ repository/gi-values-guide.pdf.

County of Wellington. 2013. *Green Legacy Program and Trees for Mapleton*. http://www. wellington.ca/en/discover/treesformapleton.asp.

Dufferin-South Simcoe Lake Stewardship Network. 2011. *Lake Simcoe Rural and Community Stewardship Program Final Report*. Prepared by John Osmok, Ministry of Natural Resources, 15 May.

Ferguson, E. 2012. "The Many Faces of Resilience: Rural Ontario Case Studies of Response to Climate Change and Peak Oil." Unpublished paper, University of Guelph, 11 May.

Green Infrastructure Ontario Coalition. 2012. *Health, Prosperity, and Sustainability: The Case for Green Infrastructure in Ontario.*

Greenbelt Foundation. 2008. *Ontario's Wealth, Canada's Future: Appreciating the Value of the Greenbelt's Eco-Services*. David Suzuki Foundation. http://www.davidsuzuki.org/publications/downloads/2008/DSF-Greenbelt-web.pdf.

Ontario Ministry of Agriculture, Food and Rural Affairs (OMAFRA). 2013. *Environment Farm Plan*. http://www.omafra.gov.on.ca/english/environment/efp/efp.htm.

Ontario Soil and Crop Improvement Association. 2013. *Canada-Ontario Farm Stewardship Program (COFSP) Environmental Cost-Share Opportunities for Farmers Available through the Canada-Ontario Environmental Farm Plan*. http://www.ontariosoilcrop.org.

Prairie Research Associates. 2011. "Environmental Farm Plans: Measuring Performance, Improvement Effectiveness, and Increasing Participation." Presentation, 15 November.

Rittel, H.W.J., and M.M. Webber. 1973. "Dilemmas in a General Theory of Planning." *Policy Sciences* 4: 155–69.

Town of Caledon. 2006. *Community Green Fund Showcase Report*. http://www.caledon.ca/en/townhall/resources/Greem_Community_Green_Fund_Showcase_Report.pdf.

Town of Mapleton. http://www.mapleton.ca/trees-mapleton.html.

CHAPTER FOUR

INCREASING ENERGY COSTS, CLIMATE CHANGE, AND PERSONAL TRANSPORTATION IN RURAL ONTARIO

ERIC MARR

In the majority of cases, living in rural Ontario means using a personal vehicle as the primary mode of transportation, and, for those of us who can, driving is an essential aspect of our daily lives that we largely take for granted. We drive to work or school, to get groceries and other supplies, to see friends and family, to seek out health care and other social services, to attend council meetings, and to vote. However, what would happen if we could not drive anymore, perhaps because of physical or financial reasons? Now consider what would happen if a large, and ever-growing, proportion of rural residents could no longer drive; how would they conduct their everyday lives? This is a question that municipalities in rural Ontario need to ask themselves.

Indeed, a direct link exists between increasing costs of oil and increasing costs of vehicle fuel. Already we see the rapidly increasing cost of fuel in Ontario, with the average cost of regular unleaded gasoline more than doubling in the past fifteen years, having increased from 58.3 cents per litre in 1996 to more than 120 cents per litre in 2011 (Statscan 2011). Now consider a similar increase going forward or even an increase at a faster rate. It will bring financial strains and transportation limitations to every segment of Ontario, but for practical reasons rural residents will be particularly impacted.

Parallel to the issues for personal transportation associated with increasing fuel prices is the issue of climate change. Concern about overall greenhouse gas emissions has moved to the forefront in recent years with increased research and societal recognition of climate change and its causes. Indeed, the transportation sector is a major contributor to greenhouse gas emissions because the internal combustion engine releases airborne pollutants such as carbon dioxide, lead, carbon monoxide, nitrogen oxides, volatile organic compounds (VOCs), particulates, and sulphur dioxide (Cullinane and Stokes 1998). This chapter proposes that the solution to the transportation conundrum posed by both climate change and increasing fuel prices can be found in public transportation.

THE CURRENT STATUS OF PERSONAL TRANSPORTATION IN RURAL ONTARIO

For anyone having lived in, or even visited, rural Ontario, it should come as no surprise that automobile use is essential to the everyday lives of residents. Indeed, rural residents in Ontario generally lack alternative forms of transportation, and personal vehicles remain the only option. Although walking or bicycling is an option for some residents living in or close to rural towns, for the many rural residents in the countryside this is not an option. It is also not an option for those with physical mobility restrictions because of age or disability.

Furthermore, by definition, rural areas have low population densities and long distances between villages or towns (Reimer and Bollman 2010). Rural residents must travel longer distances than their urban counterparts, thereby using more fuel to access services to meet equivalent needs. This need to travel greater distances has been further compounded by recent trends in centralizing services (health services, education, groceries, etc.) as opposed to previous distribution across rural communities. Once again this results in rural residents travelling increasing distances to access basic needs.

Another consideration for fuel use in rural areas is the prevalence of vehicles built for utility as opposed to fuel efficiency. Particularly in agricultural areas in Ontario, the choice of personal vehicle cannot always be based on fuel efficiency because some farming practices require certain vehicles, such as large trucks. This creates an additional disadvantage for rural residents in terms of fuel affordability.

As identified in the Canadian Senate report entitled *Beyond Freefall: Halting Rural Poverty*, a lack of access to transportation is already a problem in rural areas, particularly for those living in poverty who cannot afford to purchase and operate a personal vehicle (Standing Senate Committee on Agriculture and Forestry 2008). As fuel prices rise, this demographic can be expected to swell as more individuals cannot afford to operate personal vehicles and will require some form of alternative transportation. Similarly, as more rural residents are "priced out" of owning and operating automobiles, we can expect to see an increase in rural residents classified as low income, particularly under the Market Basket Measure (MBM), which includes transportation costs when calculating the cost of living (Reimer and Bollman 2010).

Along with the social implications of high vehicle use are the environmental impacts, including the contribution of personal vehicles to climate change. Within the rural context, there is some disagreement over the rural contribution of personal vehicle use to greenhouse gas emissions; overall, rural areas have smaller populations than urban areas and thereby, in absolute terms, contribute less in terms of greenhouse gas emissions. Rural areas also generally do not experience the same level of congestion as urban areas, which should result in fewer emissions since vehicles use less fuel per kilometre in uncongested conditions (Mees 2010). However, in per capita terms, there are questions about the contribution of rural residents. For instance, Cullinane and Stokes (1998, 134) found that, "contrary to popular assumptions, vehicle use in rural areas does contribute considerable emissions on a per capita basis. In at least one study from the UK it was found that on a per capita basis rural residents actually contributed more emissions from transport than those in urban or intermediate areas with almost all of this resulting from personal vehicle use." They state that, "while rural residents do not suffer the same congestion as their urban counterparts, they do tend to drive long distances at speeds above the optimum speed for car travel." Additional findings from the United Kingdom indicate that "people residing in rural areas on average produce nearly 50% more CO_2 from travelling than the national average" and that they were "identified as those with the highest CO_2 per person per year, in particular for travel by car (as driver), which accounted for just under 2,000kg CO_2 per year" (CRC 2008, 18, 21). Therefore, there can be a role for public transportation to play in transporting rural residents and thereby reducing greenhouse gas emissions.

THE FUTURE OF PERSONAL TRANSPORTATION IN RURAL ONTARIO

It can be argued that households in the most isolated areas, who are structurally dependent on the car (as opposed to merely reliant on it), who have a high annual mileage, and who may be among middle to low income groups could struggle to absorb the additional cost [of fuel] in the short to medium term, and face a reduced quality of life in the long term. This would involve enduring greater financial hardship, or disposing of a necessary vehicle with associated concerns for mobility, employment, housing and overall quality of life. (Gray 2000, 235)

Overall, a confluence of factors has created a situation in which vehicle reliance and fuel consumption are high in rural areas, with limited opportunities for residents to adjust travel behaviour. Now consider the effects of increasing fuel costs within this context. Indeed, we can expect that rural residents will be particularly impacted in their transportation capabilities compared with their urban counterparts.

Findings from the United Kingdom reinforce the impacts that increasing fuel prices have on rural communities, one source stating that "it has become accepted wisdom that rising fuel prices will have a devastating effect on rural communities" (Gray 2000, 233). Indeed, in a similar vein, Gray et al. (2001) investigated the taxation system known as the *fuel duty escalator,* which increased the cost of fuel, through taxation, incrementally over time.

Their findings indicate that this increasing fuel cost will significantly impact the affordability of transportation in rural areas, particularly for low-income households already making sacrifices to own and operate vehicles.

One possible means to mitigate the increasing costs of petroleum-based fuels is a replacement of fuel type. Alternative fuel sources such as biofuels, hydrogen, and electricity might be alternatives to our current reliance on petroleum. However, the German Society for Technical Cooperation (GTZ) indicates in its *International Fuel Prices 2009* report that "other types of fuels such as biofuels and hydrogen are not yet available on a substantial scale and are often even less desirable due to their economical, ecological and social consequences." Therefore, it concludes that "the transport sector must continue to focus on the reduction of travel needs, the preservation and expansion of public and non-motorised transport (walking and cycling) and the improvement of the efficiency of existing public and private transport" (GTZ 2009, 1).

Further still, within the US context, it has been estimated that, under normal replacement rates, it would take between ten and fifteen years and $1.3 trillion to replace half of the country's automobiles (Hirsch, Bezdek, and Wendling 2005). This suggests that, even if petroleum-operated automobiles were taken off the market today, it could take up to fifteen years before the stock is replaced. Therefore, regardless of whether alternative fuels and their associated infrastructures become viable replacements for petroleum in the future, transportation alternatives need to be available for the transitional period.

Perhaps the preferable method for mitigating the increased cost of fuel is an adjustment in travel behaviour and mode. As mentioned earlier, much of the travel behaviour of rural residents is based on factors beyond their control (e.g., physical distance), so the opportunity for mitigation by this means is limited. Although making less frequent trips or carpooling can sometimes be an option, electing not to drive at all likely would not. Therefore, an adjustment of travel mode remains the most likely candidate, with the most realistic possibility being some form of public transportation.

THE ROLE OF PUBLIC TRANSPORTATION

As the price of energy continues to rise and awareness of personal vehicle contribution to climate change increases, we can anticipate an expanded role for public transportation in rural Ontario. As presented here, transportation alternatives to personal vehicles can ensure that rural residents can continue to access the services and activities required for a good quality of life. Similarly, the need to mitigate climate change through reduced greenhouse gas emissions from transportation is increasingly urgent. Furthermore, the obstacles generally associated with public transportation in rural areas can be overcome.

Figure 4.1. Rural bus stop in Norway

Figure 4.2. Rural public transportation in Norway

Although a need for public transportation already exists for some demographics in rural Ontario, we can expect that this need will grow significantly as the cost of fuel makes driving prohibitively expensive for more people. Indeed, transporting people with public transit is much more efficient than doing so with personal vehicles and thereby much more cost effective for riders and communities. For instance, public transportation has been found to utilize, on average, half the energy of moving a person the same distance by personal vehicle (Shapiro, Hassett, and Arnold 2007). The same report identified additional benefits of public transportation, particularly through the lens of climate change, stating that "travel by public transportation produces, on average, 95 percent less carbon monoxide, 90 percent less volatile organic compounds, and about 45 percent less carbon dioxide and nitrogen oxide, per passenger mile, as travel by private vehicles" (9).

Along with an increase in the need for public transportation, we can also expect to see an increase in the demand for the service and thereby its feasibility. The feasibility of public transportation in rural Ontario is largely hindered by cost recovery difficulties and limited demand stemming from large distances, low population densities, and in turn inconvenient service. As the cost of energy rises, we can expect to see significant shifts toward an increase in the feasibility of the service.

In terms of demand, a shift in individual cost-benefit considerations can be expected to occur. For instance, the inconvenience related to waiting for and riding buses might become less of a consideration if the alternative is considerably more expensive. As well, because of the nature of public transportation, an increase in ridership generally creates a "virtuous circle," whereby increased ridership begets improved service that in turn produces increased ridership.

Various sources confirm that increased fuel costs do result in increased ridership in public transportation systems, including in rural and small-town settings (Mattson 2008). These sources also tend to identify some degree of lag between the increase in fuel cost and the increase in ridership, thereby suggesting that it takes time for individuals to adjust their travel behaviours; so, while demand for public transit will increase, the service will not necessarily be immediate. Further still, these studies were conducted within the context of fuel increases far below what can be expected to occur in the future, thus further confirming the anticipation of increased demand for public transit.

In connection with the increased demand, we can expect an increase in the feasibility of public transportation services in rural Ontario. Indeed, beyond the expectation of an increase in ridership justifying routes with comparably limited population densities, we can also expect that services will be able to recover more costs. Through economies of scale, the marginal cost of additional passengers using a transportation system is quite low. Therefore, any increase in use will result in an increase in revenue and little extra cost of operation.

Nevertheless, assuming that the public transportation system utilizes petroleum-based fuels, the cost of operation will increase as the cost of energy rises. This does pose a problem, for these systems are already generally viewed as being prohibitively expensive for rural municipalities in Ontario. However, inputs can be expected to increase because of the heightened demand for services, which might offset any increase in operating costs. As well, we can expect that users will be willing to pay increasingly more for fares since any increase in fuel cost will be felt more directly by automobile drivers than when split between all riders in more fuel-efficient public transportation services.

In summary, for a variety of interconnected reasons, we can expect that the increasing cost of energy will directly result in higher costs for personal transportation in rural Ontario, with impacts felt more strongly there than in urban areas. As more and more rural residents are unable to afford the cost of driving, alternatives will need to be in place to support this growing segment of the population. One such alternative is a public transportation system that will likely increase in feasibility as the cost of fuel rises while also providing opportunities to mitigate greenhouse gas emissions.

Although there are good reasons why most rural municipalities in Ontario have not yet created transportation systems, there are several operating systems in rural contexts across the globe and even some from Canada that can be used as potential models. As well, rural municipalities cannot afford the gamble of hoping that the technology and infrastructure of alternative fuels will be available in time. Instead, rural municipalities in Ontario should consider their options and start small, with opportunities to expand, because the capacity to operate a successful public transportation system must be fostered over time, not when it is too late.

References

Commission for Rural Communities (UK). 2008. "Thinking about Rural Transport: Contribution to Sustainable Rural Communities." *CRC WEB 30.* Cheltenham, UK: Commission for Rural Communities (CRC) and Transport Research Laboratory (TRL).

Cullinane, S., and G. Stokes. 1998. *Rural Transport Policy.* Amsterdam: Pergamon.

Gray, D. 2000. "Does Scotland Have a Rural Transport Problem?" Paper presented at the European Transport Conference, Cambridge, UK, 11–12 September.

Gray, D., J. Farrington, J. Shaw, S. Martin, and D. Roberts. 2001. "Car Dependence in Rural Scotland: Transport Policy, Devolution, and the Impact of the Fuel Duty Escalator." *Journal of Rural Studies* 17: 113–25.

GTZ. 2009. "International Fuel Prices 2009—6th Edition." Deutsche Gesellschaft für Technische Zusammenarbeit (GTZ). http://www.gtz.de/de/dokumente/gtz2009-en-ifp-full-version.pdf.

Hirsch, R.L., R. Bezdek, and R. Wendling. 2005. "Peaking of World Oil Production: Impacts, Mitigation, and Risk Management." National Energy Technology Laboratory (NETL). http://www.netl.doe.gov/publications/others/pdf/oil_peaking_netl.pdf.

Mattson, J. 2008. "Effects of Rising Gas Prices on Bus Ridership for Small Urban and Rural Transit Systems." Small Urban and Rural Transit Center, Upper Great Plains Transportation Institute, North Dakota State University.

Mees, P. 2010. *Transport for Suburbia: Beyond the Automobile Age.* London: Earthscan.

Reimer, B., and R.D. Bollman. 2010. "Understanding Rural Canada: Implications for Rural Development Policy and Rural Planning Policy." In *Rural Planning and Development in Canada,* edited by D. Douglas, 10–52. Toronto: Nelson.

Shapiro, R.J., K.A. Hassett, and F.S. Arnold. 2007. "Conserving Energy and Preserving the Environment: The Role of Public Transportation." American Public Transportation Association.

Standing Senate Committee on Agriculture and Forestry. 2008. "Beyond Freefall: Halting Rural Poverty." Senate of Canada. http://www.parl.gc.ca/39/2/parlbus/commbus/senate/com-e/agri-e/rep-e/rep09jun08-e.pdf.

Statscan. 2011. "Energy Statistics Handbook: Third Quarter 2010." Statistics Canada. http://www.statcan.gc.ca/pub/57-601-x/57-601-x2010003-eng.pdf.

CHAPTER FIVE

EDEN MILLS GOING CARBON NEUTRAL: CASE STUDY OF A COMMUNITY APPROACH

ÉMANUÈLE LAPIERRE-FORTIN, WAYNE CALDWELL,
AND JOHN F. DEVLIN, WITH CHRIS WHITE

This chapter presents findings from research that we conducted in 2010 and 2011 on the question of community resilience to climate change and rising oil prices in two rural communities in southern Ontario, centred on the following research question: *how are community organizations responding to climate change and energy uncertainty, and how could their response be strengthened?* A definition and a conceptual framework on community resilience borrowed from Magis (2007a, 2007b, 2009, 2010) are presented, in light of which highlights and insights are collected from the first case study, Eden Mills Going Carbon Neutral. The second case study, Transition Guelph, is the focus of the following chapter and uses the same framework.

BACKGROUND ON COMMUNITY RESILIENCE

Planners and citizens' organizations can engage in a process of deliberate transformation to reach the outcome of "adaptedness"[1] in response to system disturbances, such as climate change and energy uncertainty. In so doing, they can build community resilience. Magis (2009, 12) concludes that "the most appropriate response to a system's disruption will vary from maintenance to adaptation to transformation"; transformation, which enables new structures and processes to occur, is a healthy response to enable survival of the system if disturbances push it to thresholds at which minor adaptations are not sufficient.

Magis (2010, 401) proposes a general definition of community[2] resilience that aligns with the interpretation of resilience building as "initiating controlled positive change": "Community resilience is the existence, development, and engagement of community resources by community members to thrive in an environment characterized by change, uncertainty, unpredictability, and surprise. Resilient communities intentionally develop personal and collective capacity to respond to and influence change, to sustain and renew the community and to develop new trajectories for the community's future." This definition of resilience focuses on the agency of community members and groups, or the *active agents*, to influence the course of change even when the change is unpredictable, as is the case with climate change and energy uncertainty, as well as to plan and act strategically to create a desirable future. This definition is suitable to analyze the processes put in place by community organizations to build resilience by engaging in deliberate transformation in pursuit of adaptedness. According to Berkes and Sexas (2005, in Magis 2009, 8), to be resilient, "communities need to learn to live with change and uncertainty, and actively build the capacity to thrive in that context." As a result, this research focuses on the processes established by communities to make proactive changes in the face of long-term trends such as climate change and energy uncertainty.

Magis (2007a, 2007b, 2009, 2010) has developed a community resilience framework that is an important landmark in the community resilience literature and is used to inform this case study and the Transition Guelph case study. The framework is depicted in Figure 5.1.

This framework can be used as a measurement and assessment tool in any community to stimulate better discussion among stakeholders, thus contributing to and focusing on the process of community change (Kelly 2010) through the mobilization of community resources or community capital. Community capital is the linchpin of a community's assets and resources and is understood here as the cumulative forms of natural, cultural, human, social, political, financial, and built capital (Emery and Flora 2006; Flora et al. n.d.; Flora and Flora 2004, in Magis 2007a).[3]

The right-hand side of the diagram relates to the ways in which action to build community resilience is undertaken. *Active agents* refer to the dominant ideas in this chapter that, "although external forces impact the community, the community can influence its well-being and take a leadership role in doing so" (Magis 2010, 411) and that community members are the primary and active agents in the community's resilience, even though they cannot control all of the conditions that affect their community (Magis 2009). Active agents can include citizens' organizations and local governments seeking to build resilience to climate change and energy uncertainty. The next dimension, *collective action*, suggests that strategies are more efficacious when people from diverse and autonomous groups work together. Magis (2010, 411) adds that "the extraordinary work of a singular individual or group of individuals is insufficient." Next she points to the importance of *strategic action* for community resilience; actions need to be in line with the community's visions and objectives, prioritized, planned, deliberated, and implemented.

On the left-hand side of the diagram are concepts relating to *community resources* or capital. When speaking of community resources, Magis (2010)

COMMUNITY RESOURCE/ CAPITAL	DESCRIPTION
NATURAL CAPITAL	Resources and ecosystem services from the natural world
HUMAN CAPITAL	Individuals' innate and acquired attributes (capacities, skills, knowledge, health, leadership)
CULTURAL CAPITAL	Communities' ways of knowing the world, their values and assumptions; manifested with symbols in language, art, and customs
FINANCIAL CAPITAL	Financial resources available to be invested in the community
BUILT CAPITAL	Community's physical assets and built infrastructure
POLITICAL CAPITAL	Community members' ability to access resources, power, and power brokers and to impact the rules and regulations that affect the community
SOCIAL CAPITAL	Ability and willingness of community members to participate in actions directed to community objectives and to the process of engagement: that is, individuals acting alone and collectively in community organizations, groups, and networks. Putnam (1995) describes "bonding" social capital as social networks between socially homogeneous groups of people and "bridging" social capital as social networks between socially heterogeneous groups of people. Bridging social capital between groups of people who are not at the same level of authority (e.g., between citizens and local governments) is sometimes referred to as "linking" social capital.

Table 5.1. Types of community resources. Source: Magis (2010)

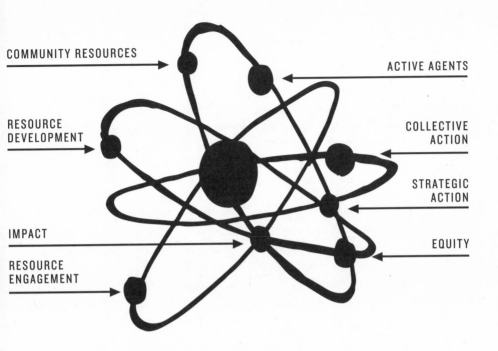

COMMUNITY RESOURCES

ACTIVE AGENTS

RESOURCE
DEVELOPMENT

COLLECTIVE
ACTION

STRATEGIC
ACTION

IMPACT

EQUITY

RESOURCE
ENGAGEMENT

Figure 5.2. Conceptual framework for building community resilience

Figure 5.2. North America's first carbon-neutral village

refers to and describes seven types of community capital (natural, human, cultural, financial, built, political, and social) that need to be present, developed, and engaged in resilient communities. They are summarized in Table 5.1.

Magis (personal communication, 23 March 2011) stresses the importance of analyzing how these resources, present in most communities, are actually used to build their capacity to adapt to change. In fact, in Figure 5.1, the term *community resources* refers to the existence in the community of the seven aforementioned types of capital, while the *development* of community resources refers to the dynamic character of those resources, which can be expanded or depleted through community action. Magis (2010, 410) distinguishes between the community resources and the development of those resources by noting that "developing community resilience requires action taken, not simply the capacity to act." The following dimension of community resilience is the *engagement* of community resources, which refers to developing buy-in of a shared community objective and mobilizing the community's resources toward fulfilling that objective. This dimension also incorporates public participation.

The last two dimensions of community resilience retained by Magis are *equity*, which refers to social justice, equality, and intergenerational distributional concerns, and *impact*, evidenced by the community's level of success in implementing plans and adapting to change.

The three variables selected for case study analysis in this chapter and the following chapter include (1) development of bridging social capital, (2) development of linking social capital, and (3) development of human capital.[4] These case studies were developed from literature and document reviews, participant observations, semi-structured interviews with members of eighteen organizations, and eleven key informant interviews with community members such as mayors, members of Parliament, municipal staff members, and partner organizations.

CASE STUDY DESCRIPTION

Eden Mills Going Carbon Neutral (EMGCN) is located in Eden Mills, a village of 350 people located thirteen kilometres from Guelph in southern Ontario. The EMGCN initiative seeks to have Eden Mills become the first carbon-neutral village in North America by producing no more carbon dioxide than it absorbs within its boundaries. To do so, it has identified a three-pronged approach of reducing, replacing, and absorbing emissions. In collaboration with the University of Guelph to measure change, Eden Mills has conducted three annual surveys to calculate the household and travel carbon emissions of village residents. To assist residents in taking action, the village has conducted many practical workshops on energy efficiency in the home.

EMGCN started when residents of Eden Mills heard about and visited Ashton Hayes, England, seeking to become the first carbon-neutral village in that country. After an informal gathering of neighbours in November 2007 to discuss how Eden Mills could also go carbon-neutral, a core team of roughly ten people took four months to organize the official "launch" of the program. During that time, they raised $3,000 (much of it as in-kind donations), got a resolution of support passed unanimously at the township level, secured free space from and collaboration with the Community Club, were taken under the umbrella of the Millpond Association (which has charitable status), hand-delivered a first newsletter to all households in Eden Mills, and set up communications. Every household was invited to come to the launch in November 2007, a standing-room-only event with local and national media coverage. The event covered the basics of global warming and carbon neutrality and was celebratory in nature, including free food, drinks, and live music, and featuring guest addresses from local politicians and video messages from Ashton Hayes Going Carbon Neutral. The launch, a call to action based on concrete work in progress, highlighted several projects and collaborations already under way and asked for volunteers to work on specific aspects of the project.

A household survey is used to measure the carbon footprint of the village every two years, based on confidential, individual home results that are also provided to residents for their own purposes. Since 50 percent of carbon emissions were absorbed by the plants and trees of the village, the carbon emissions remaining to be neutralized amounted to twelve tonnes of carbon per household in 2007.[5] The survey had a response rate of 69 percent in year three, and its results indicated that "by 2009 the village footprint had decreased from the 2007 baseline by 6%" (Sword 2010, 20). By 2013, the carbon footprint had decreased by 36 percent from the 2007 level (Laing, personal communication, 13 April 2013).

To achieve the goal of becoming carbon neutral, EMGCN has a close focus on education. The Eden Mills Youth Group (Grades 7 to 12) is a unique component of the organization. It meets twice a month in the evening and once a month on the weekend to engage in outdoor activities, grow a community vegetable garden in the centre of the village, and help to organize an Eco Conference for Grade 9 students from different schools to brainstorm ways to reduce an individual's carbon footprint. It also had a major role in organizing Earth Day, during which, over three years, more than 5,000 trees were planted by Eden Mills residents to reduce the carbon footprint of the village.

In addition, EMGCN got a Communities Go Green Grant to support communications. The grant enabled the group to produce two practical handbooks as references for other communities and eleven DVDs of videotaped workshops on topics such as geothermal heating, solar hot water, photovoltaic arrays, organic food growing and preserving, and home insulation.

In the past two years, the group has focused on "greening" the Community Hall as a demonstration project. The hall is the sole community space in the village and its oldest building at 150 years. The group trusted that retrofitting the least environmentally friendly building in the village would enable it to lead by example while making sure that the hall could still be used by the community even if energy costs were to rise. Funds raised for this project totalled $400,000 and were secured from the Ontario Trillium Foundation as well as donations. Retrofits include a new metal roof, replacement of all windows by three-pane, highly efficient windows, and reinsulation throughout the building, all of which reduced the energy cost by almost 40 percent. The energy system is currently being changed so that it is not reliant on fossil fuels (a heat pump and air-recovery units will be installed). The cost of fully insulating the hall is estimated to be $225,000; the group is still fundraising to finalize the project.

Finally, the EMGCN group also had a joint project of installing a photovoltaic array with the Community Futures Development Corporation. Such systems were increasingly being installed in the village prior to cancellation of the Community Feed-In Tariff Program.

The *So You Want to Go Carbon Neutral? It Takes a Village!* handbook (Sword 2010) is recommended reading for anyone who would like to find out more about EMGCN, along with the group's website at http://goingcarbonneutral.ca/.

INTERVIEW FINDINGS

According to interviewees, the most significant change has been increased awareness of the problem since residents have done things collectively, talked to people, encouraged them to take action, and worked with children. A high rate of participation was noted, demonstrated by a high survey response rate and attendance of between forty and fifty people at workshops and meetings. One interviewee said that, "after three and a half years, the group is still in place, has a good reputation, and people are proud to live in a village that wants to be the first in North America to become carbon neutral." Furthermore, residents have made and continue to make significant changes in their households. Several have replaced heating systems, and one home was completely retrofitted for energy efficiency and sustainability using technology such as a composting toilet and a highly efficient heat pump. These improvements have inspired others to follow suit.

EMGCN has been referred to by participants as providing a "package deal" comprising not only environmental activities but also fun opportunities to get together in a relaxed way that stimulates community spirit. Examples include the celebrations after the survey results are out and the party that marked the village's first solar array hook-up to the grid. Community celebrations are successful in promoting change and in reaching out to people who might not be drawn instinctively to making behavioural changes.

Figure 5.3. First solar panels on Community Hall installed in 2010

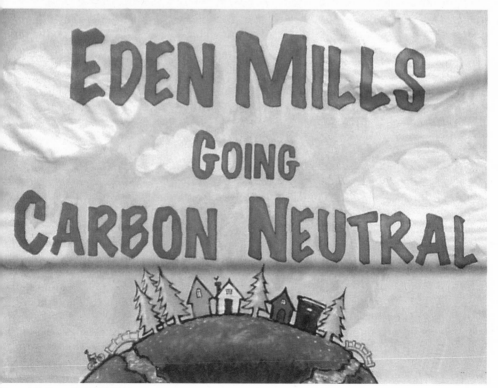

Figure 5.4. Community Hall banner

In terms of working with local governments, EMGCN has a positive working relationship with the Township of Guelph-Eramosa. The mayor sourced free trees for Eden Mills from the township, and the township passed a resolution to attain 40 percent tree coverage in rural areas and 30 percent tree coverage in urban areas in thirty years. According to Chris White, mayor of Guelph-Eramosa, the role of the township in responding to climate change and energy uncertainty is that of a partner and follower; it is the township's duty to support grassroots groups doing positive things. For EMGCN, the township can provide support by being aware of opportunities, connecting the group with upper levels of government, increasing exposure for the EMGCN model, educating citizens, leveraging grants by providing seed money, and providing stability and credibility to the group. White adds that "we need champions to drive this thing, to know the system, come up with a plan or strategy and bring it to the table" (personal communication, 19 November 2010). In fact, White invited EMGCN to make a formal presentation to the annual meeting of the Rural Ontario Municipalities Association in 2008 and noted that the township continues to learn from EMGCN.

Interestingly, the Environmental Stewardship Committee of the Guelph-Eramosa Township was started around the same time as EMGCN, and a member of EMGCN was invited to join it. As a result, the Stewardship Committee provided some funding for the school projects described below and for the signs announcing the carbon neutrality goal at the entrance to the village. On the flipside, the mayor acknowledges that a committee that meets for an hour every month is by definition slow moving.

EMGCN also worked with the township on transportation, repeatedly identified as the biggest external challenge to carbon neutrality in Eden Mills. With no public transit, the village is dependent on personal cars. As an initial step, the township's Trail Committee recently approved a trail between Eden Mills and Rockwood, where primary school children go to school. EMGCN held a Transportation Café in 2010 to get input on how to create a new system that saves money and time and reduces emissions. A follow-up questionnaire on some short-term ideas was circulated to the EMGCN listserve in early March 2011. Transportation became a priority for the group at that time, and it is still seen as the biggest challenge to carbon neutrality in Eden Mills. Since 2011, the group has met with the City of Guelph and the Township of Guelph-Eramosa to discuss its overall transportation plan, which has motivated the township to find solutions. As of early 2013, the mayor asserted that the township has met with the Guelph mayor and the school board to discuss the possibility of sharing buses. He also intended to commission a study on transportation across the County of Wellington, which would create ripple effects outside Eden Mills (White, personal communication, 2 March 2013).

In addition to the relationship with the Township, EMGCN inspired the Upper Grand District School Board on the Green Classroom program, a project involving solid waste reduction informed by social marketing research. The project has kids talking to kids about garbage and reusing it creatively. It started in Rockwood Centennial Elementary School, attended by Eden Mills children, and then the school board adopted it and made it mandatory for all its schools. Following that, a member of EMGCN created a proposal for the school board to build eco-efficient classrooms as prototypes for houses and buildings of the future to change how kids view physical spaces. The school board agreed to build nine classrooms, also known as Environmental Learning Centres (the first two were built in Orangeville, and another one was scheduled to open in Harriston in 2013), and to develop curriculum for an energy unit in Grade 5. All Grade 5 students spend the year in these hands-on classrooms, which provide material for a wide variety of new environmental teaching units experienced directly by the students and available to them anywhere through the school board website.

EMGCN has been supporting the mandate of the Green Legacy Fund of Wellington County from the outset (see Chapter 3). An EMGCN member developed a project in which schoolchildren and female inmates from the Vanier Centre for Women volunteer to plant trees, thus absorbing carbon. As part of a regular school day, kids learn about trees and plant them, all at no cost to the school. The program started with 600 kids from Grades 3, 5, and 6 who transplanted 20,000 seedlings, and in 2011 there were 6,000 kids from across the county involved in the project (every child from Kindergarten to Grade 8). According to the project organizer, female inmates are learning basic job skills and feeling useful.

The relationship with the Eden Mills Millpond Conservation Association was identified by one interviewee as being important in structuring and providing support to EMGCN as well as making fundraising and applying for grants easier. EMGCN operates as a project of the Millpond Conservation Association, a registered charity, and as a result is able to issue tax receipts for donations. In addition, the group is connected to the Eden Mills Natural World Speakers' Series of monthly speakers through cross-membership of one EMGCN coordinating team member. The two groups have common interests.

In terms of challenges, some interviewees identify outreach since community organizations cannot reach everybody through events. There is a sense that it is important to keep activities diverse and inclusive so that there is something for everyone; this also means balancing big ideas and practical actions. This is linked to the challenge of changing behaviour, which can often be addressed by providing information on how much households can save by being more energy efficient. One person mentioned that a lot of the changes promoted require cultural change in terms of what is considered "normal": staying home instead of going south for holidays, washing dishes by hand, carpooling, et cetera.

EMGCN aims to empower people so that the work comes from the ground up and not always from the same core group. This approach was mentioned by a collaborator as important for development, but it has the drawback of being less "efficient" in terms of getting things done. Also, the organization operates in a context in which there is a limited number of volunteers and many volunteer opportunities. To address that challenge and make sure that organizations are not competing against each other, EMGCN has engaged in a number of joint fundraising events. Another common challenge mentioned that seems to be addressed by EMGCN is maintaining momentum and ensuring succession. According to one participant, while many people come to events, only about twelve very loyal people come to every meeting, and the group has never been able to move that number to thirty or fifty. Issues of burnout and time and energy constraints were mentioned by one person. A newcomer to Eden Mills mentioned being quickly welcomed to the village and being invited to participate in EMGCN by core group members.

Another challenge has been that some data were lost in the second year's survey, and some of the analysis from the University of Guelph students in that year's survey was labelled as questionable by a consulting statistician. It seems that there was a lack of clarity between the EMGCN group and the university about who was responsible for supervising the students, a question that has now been addressed.

The group has learned that there is a constant need to keep the initiative fresh and visible and that it is important not to force people, to be careful how the message is pitched, and to avoid being viewed in a negative way as a sort of "carbon police."

COMMUNITY RESILIENCE ANALYSIS

Bridging Social Capital Development – EMGCN collaborates with many other organizations in working toward its mandate. As mentioned earlier, EMGCN inspired the Upper Grand District School Board on the Green Classroom program. Nine pilot classrooms were built, and a waste reduction curriculum has been delivered to all Grade 5 students in the school district. Furthermore, the group has a working relationship with the Green Legacy Fund of Wellington County, which has resulted in a mutually beneficial partnership with its tree nursery and the Grand River Conservation Authority; an EMGCN member developed projects that incorporated over 6,000 schoolchildren as well as female inmates from the Vanier Centre for Women as volunteers to plant trees. In addition, the working relationship with the Community Club is significant. Making the Community Hall (the only common space in the village) as "green" as possible has been one visible demonstration project of EMGCN. The relationship with the Eden Mills Millpond Conservation Association was identified by one interviewee as being important in structuring and providing support to the organization as well as making fundraising and applying for grants easier. EMGCN operates as a project of the Millpond Conservation Association, a registered charity, and as a result is able to issue tax receipts for donations. In addition, the group is connected to the Eden Mills Natural World Speakers' Series (monthly speakers) through cross-membership of one EMGCN coordinating team member. The two groups have common interests.

Human Capital Development – The focus on education to galvanize efforts toward the reducing, replacing, and absorbing strategy indicates the importance of human capital in EMGCN. The group's practical workshops, which invite people to improve their physical capital through house-based retrofits, are all videotaped so that they can be borrowed by anyone in the village. Also, the active Youth Group is a testimony to the role of educating future generations on issues for which they will soon be responsible. Furthermore, EMGCN, compared with other groups, has handled common challenges of volunteer organizations, namely burnout and resource scarcity, very well. This can be attributed to a local couple's decision that one parent in the household would devote him- or herself fully to the initiative instead of selling labour on the market. That decision resulted in EMGCN having a full-time volunteer for two and a half years, a rare advantage that had wide-ranging ripple effects, especially in the tree-planting and waste reduction programs with local classrooms.

Linking Social Capital Development – Key informant interviews suggested that the role of the local government in terms of building community capital for resilience is to be a facilitator and help to develop and engage resources, especially by providing financial capital. In the case of EMGCN, the township has heightened the credibility of the citizens' initiative by giving it outside exposure, which arguably has increased its impact outside Eden Mills. "Sanctioning" the initiative also helps to stimulate diversity by mainstreaming it, which can contribute to the equity aspect in Magis's framework. According to Chris White, mayor of Guelph-Eramosa, the role of the township in responding to climate change and energy uncertainty is that of a partner and follower and is comprised of being aware of opportunities, connecting the group with upper levels of government, bringing exposure to the EMGCN model, educating citizens, getting leverage when applying for grants by providing seed money, and giving stability and credibility to the group. Finally, establishment of the Environmental Stewardship Committee, which reports to council, provides a longer-term vision of resilience essential in engaging in strategic action.

CONCLUSION

The research presented in this chapter sought to answer the question *how are community organizations seeking to build community resilience to climate change and energy uncertainty in Ontario?* The findings are summarized in Table 5.2.

Given the short period of time that the organization has been in operation, it is too early to provide a definitive evaluation of its impacts on community resilience. Based on the empirical data from this research and informed by the literature on community resilience, especially the eight dimensions of the Magis framework presented above, we summarize our impressions of the synthesis of internal strengths and weaknesses and indicate some external opportunities and threats in the future in the form of a Strengths-Weaknesses-Opportunities-Threats (SWOT) analysis in Table 5.3. This information was used in a workshop session at a Community Research Report Back event held at the University of Guelph in April 2011 to guide a dialogue on concrete actions for "what's next?" It was also presented as a working tool for the group, understanding that community resilience building is a learning process that unfolds over a long period of time. Table 5.3 marks the end of this chapter; the following chapter presents concluding remarks on the EMGCN and Transition Guelph case studies.

EDEN MILLS GOING CARBON NEUTRAL

DEVELOPMENT OF BRIDGING SOCIAL CAPITAL	Healthy working relationships with other organizations (namely, Community Club, Millpond Conservation Association, University of Guelph, Transition Guelph, etc.)
DEVELOPMENT OF LINKING SOCIAL CAPITAL	Participation on Township Environmental Committee Greetings at launch by township mayor Political and financial support by township Working projects with school board (Green Classroom program) and county (through Green Legacy Fund) Instigating township action on transportation file
DEVELOPMENT OF HUMAN CAPITAL	Videotaped workshops on energy conservation Educational opportunities for youth through Youth Group activities and tree-planting project Print publications, newsletters, and website provide good information

Table 5.2. Summary of findings

SWOT ANALYSIS OF EDEN MILLS GOING CARBON NEUTRAL

STRENGTHS	Exposure to large audiences facilitates impact (TV, conferences—Canadian Institute of Planners, Rural Ontario Municipalities Association).
	The Community Hall as a demonstration project and the Green Classroom program show development of built capital through investment of time and money.
	Integration and collaboration with other organizations in Eden Mills exemplifies bridging social capital.
	The presence of a full-time volunteer for two and a half years provided significant human capital to the initiative.
	The high response rate for surveys (73 percent in year three) shows community support (resource engagement).
	The innovative Youth Group and tree-planting program for students engage them as active agents at a young age while developing natural capital.
	Celebrations of and social events for the cause of EMGCN show the engagement of cultural capital.
WEAKNESSES	Lack of uptake of recommendations of the transportation meeting for five months shows a weakness in strategic action.
	Difficulty in outreach to part of the village population shows limitations in bridging social capital common to many environmental organizations.
	Collective action is limited by the difficulty in growing the group of twelve very loyal people to thirty or fifty.
OPPORTUNITIES	Continued good external relations with the township and province exemplify linking social capital.
	Participation on the Township Environmental Committee shows engagement of human capital.
	Engagement of a statistical consultant will build human capital of the University of Guelph students undertaking the survey and analyzing its results.
	Collaboration with University of Guelph faculty on forest and tall grass carbon absorption research can further develop natural capital.
THREATS	Insurance policies could make it difficult to go ahead with the village solar array and further develop built capital.
	The end of federal and provincial incentives programs for retrofits and solar thermal systems will make developing built capital more difficult.
	Some level of conflict with municipal staff could jeopardize linking social capital.

Table 5.3. SWOT analysis of Eden Mills Going Carbon Neutral

Notes

1   Nelson, Adger, and Brown (2007) define "adaptedness" as "the status or characteristic of being adapted," and they define "adaptation" as "the decision-making process and the set of actions undertaken to maintain the capacity to deal with current or future predicted change."

2   Magis (2007b, 2) defines community "as a social grouping of people residing in a specific geographic territory. The community has a particular history, specific demographic patterns and houses, industries and organizations. Community members establish patterns of interaction for multiple purposes, e.g., political, economic and social, and can mobilize community resources to take collective action for the benefit of individuals and/or the community. While towns are typical communities, rural communities can extend beyond the city limits or may be unincorporated and larger cities may be comprised of several smaller communities."

3   In this chapter, developing "linking" social capital between citizens and local governments is understood as synonymous with developing political capital.

4   More information on the rationale for these choices is available in Lapierre-Fortin (2011). Field research and analysis were undertaken in 2011, and the case study descriptions were updated in 2013.

5   University of Guelph students are currently developing a Made in Eden Mills carbon offset program.

References

Emery, Mary, and Cornelia Flora. 2006. "Spiraling-Up: Mapping Community Transformation with Community Capitals Framework." *Journal of the Community Development Society* 37 (1): 19–35.

Kelly, T.J. 2010. "Five Simple Rules for Evaluating Complex Community Initiatives." Annie Casey Foundation. http://www.frbsf.org/publications/community/ investments/1005/T_Kelly.pdf.

Lapierre-Fortin, É. 2011. "Weathering the Perfect Storm: How Two Citizens' Groups Are Building Resilience to Climate Change and Peak Oil in Ontario." MSc rural planning and development major research paper, University of Guelph.

Magis, K. 2007a. "Community Resilience Literature and Practice Review." US Roundtable on Sustainable Forests, September, Special Session on Indicator 38: Community Resilience. http://www.sustainableforests.net/summaries.php.

——. 2007b. "Community Resilience Literature and Practice Review: Executive Summary." US Roundtable on Sustainable Forests, September, Special Session on Indicator 38: Community Resilience.

——. 2009. "Indicator 38: The Resilience of Forest-Based Communities." White paper for the 2010 National Report on the Sustainability of the United States Forests. Washington, DC: United States Department of Agriculture, Forest Service.

——. 2010. "Community Resilience: An Indicator of Social Sustainability." *Society and Natural Resources* 23: 401–16.

Nelson, D.R., W.N. Adger, and K. Brown. 2007. "Adaptation to Environmental Change: Contributions of a Resilience Framework." *Annual Review of Environment and Resources* 32, 1: 395–419.

Putnam, R.D. 1995. "Bowling Alone: America's Declining Social Capital." *Journal of Democracy* 6, 1: 65–78.

Sword, L. 2010. *So, You Want to Go Carbon Neutral? It Takes a Village!* Eden Mills, ON: EMGCN.

CHAPTER SIX

THE TRANSITION RESPONSE: PRACTICAL RESILIENCE STRATEGIES

ÉMANUÈLE LAPIERRE-FORTIN, WAYNE CALDWELL, JOHN F. DEVLIN, AND SALLY LUDWIG

This chapter is the outcome of collaboration between the University of Guelph and Transition Guelph. Transition Guelph is the second case studied by Lapierre-Fortin in 2011, and more details on the research are presented in Chapter 5. This chapter offers a snapshot of the transition movement in Guelph in 2010–11 along with an update on its evolution from the perspective of 2013 with Sally Ludwig of the Transition Guelph Steering Committee.

BACKGROUND: THE TRANSITION MOVEMENT

Transition Towns is an international network of communities seeking to build local resilience to climate change, resource depletion, rising energy prices, and economic instability and inequity. The idea of Transition Towns came to the founder, Rob Hopkins, a permaculture[1] instructor, when he asked his students to create an Energy Descent Action Plan for the town of Kinsale, Ireland. Upon his move to Totnes, England, Hopkins and several colleagues started to develop a model for volunteer-led community resilience based largely on relocalization; he shared the model in *The Transition Handbook: From Oil Dependency to Local Resilience* (2008). Hopkins articulated the principles that underpin the model: a vision of the desired outcome, inclusion and dialogue, focus on raising awareness in a context of mixed messages from the media, rebuilding resilience, psychological insights, and credible solutions at an appropriate, community-level scale.

Hopkins and others in Transition Network Ltd., a non-profit organization based in Totnes, went on to develop a "Collection of Ideas and Solutions for Setting Up and Running a Successful Transition Initiative" on the network website (http://transitionnetwork.org/patterns) and publish a second iteration in *The Transition Companion: Making Your Community More Resilient in Uncertain Times* (Hopkins 2011). It describes current community practices as six sets of patterns, each comprised of several tools and ingredients. The sets of patterns are organized as follows:

starting out: tools for thinking, communicating, and visioning, group and personal skills, and activities that appear to help in the early stage of forming a transition initiative;

deepening: maintaining the initiative beyond the start-up stage (designing to sustain the organization and deepen the work);

connecting: engaging community members, collaborators, and supporters;

building: scaling up the initiative; and

daring to dream: creating a groundswell of change (policies and community projects for a relocalized economy). (ibid.)

These sets of patterns are a new way of communicating transition. They have replaced the "Twelve Steps of Transition" detailed in the first edition of *The Transition Handbook* in an effort to prevent transition initiatives from seeking to follow the twelve steps chronologically at the expense of creativity and innovation in response to local conditions (ibid.).

The connecting pattern includes ingredients such as forming networks, involving the council, working with local businesses, and engaging young people. They indicate the importance of developing social capital through collaboration with other organizations, community residents, and the local government in the transition model.

Haxeltine and Seyfang (2009, 6) offer a description of the sectors in which the transition movement is typically active:

While each Transition initiative develops its own action plans and priorities, there are common themes and activities which have developed throughout the movement. Among these are: local energy generation; local food production; farmers markets; community gardening and composting; designing and building eco-housing; local currencies; personal development work; skill-sharing and education; recycling and repair schemes; car-sharing, and promoting cycling; supporting energy demand-reduction through self-help clubs, and so on. Though each activity might appear piecemeal and not particularly radical, a deeper examination reveals that they all aim to offer some aspect of an alternative set of systems of provision, based around deeper green values and a rejection of consumerism.

This type of activity is championed not only by transition initiatives; however, what makes them special is the framing of their activities using the language of "resilience to climate change and peak oil." Haxeltine and Seyfang (ibid., 8) add that "the role of the Transition movement is seen [by active members of transition initiatives] as being to build a well-developed, innovative niche of resilient sustainability, to be ready to compete with the regime as it dies away, and so to avoid the (socially, ecologically and economically) less-desirable scenarios."

Community resilience is seen in transition as a chance to develop new economic and social paradigms, filled with alternative business models and an emphasis on learning practical skills for self-sufficiency and a reduced environmental footprint. This is understood as leading to a community that is healthier, happier, more adaptable, and less vulnerable to risk and uncertainty (Colussi and Rowcliff 2000; Hopkins 2011). This includes training young people in a wide range of skills; participatory planning, budgeting, and decision making; community ownership of assets; access to agricultural land; and localization of diversified economic activity to meet the needs of the population. The movement suggests that communities use measurable indicators of resilience such as carbon emissions; percentage of food consumed locally that was produced within a given radius; percentage of essential goods manufactured and energy produced by local sources; ratio of car-parking space to productive land use; number of locally owned businesses; and proportion of community employed locally; the movement also uses more informal criteria such as number of teenage residents competent to grow different vegetables and degree of engagement in practical transition work by local citizens (Hopkins 2008).

Also unique is the transition model's use of principles of permaculture, a design system that develops sustainable human living environments (ibid.). Creation of regenerative systems is possible, permaculturists believe, when designers use patterns modelled on those found in natural ecosystems. Useful connections among elements as well as synergies in the sum of the parts of the system (people, groups, and institutions in human cultural systems) are understood to be key.

TRANSITION GUELPH

Transition Guelph (TG) started in September 2008 when three people, who had independently signed up on the Transition Network website as being interested in the model, got together to discuss starting a local transition initiative. Its public presence started with a public meeting in December at which those present created a steering group. According to the co-founders, there was high attendance at initial meetings and a lot of interest within the community, and the group was soon struggling to keep up with this interest. People were eager to volunteer and join a project; they were less prepared, it seemed, to design a project from the ground up. In its first two years of existence, TG focused largely on raising awareness and organizing many film screenings and speakers events, including an Energy Fair co-sponsored in 2009 with the Green Team of St. James Apostle Anglican Church, a conference on peak oil called Our Environmental Future co-sponsored with the Council of Canadians and a Guelph city councillor, and a presentation by author and filmmaker Patrick Murphy. TG volunteers also began to facilitate reskilling workshops, teaching practical skills for relocalization and self-reliance, which were received positively.

Since 2008, TG has held regular and well-attended general meetings (often between twenty and fifty people attend), and it has formed working groups with a focus on cultural change; they are active in the areas of local food, intentional communities, heart and soul (or "inner transition"), education, health, time banking, localizing the economy, and fair trade. Participants at general, steering, and working group meetings have engaged in visioning activities that have evolved into a series of ten vision documents on community resilience in various sectors, such as business, food, education, energy use, and transportation. They have participated in creating a "transition timeline" that envisions what a Resilient Guelph might look like in 2030 and "backcast" the steps required to get there. Together the timeline and vision documents set the foundation for a Resilience Action Plan, a strategic plan for intentional relocalization of the community as a whole that the group continues to develop.

TG volunteers also created an evaluation plan with proxy measures of community resilience relating to individuals' carbon footprints and use of gasoline and incorporated it into the 2010 membership survey, sent to the 500-member distribution list. In the following year, they also created a Household Resilience Assessment, available on the TG website (www.transitionguelph.org), for self-evaluating the resilience of individual households.

After months of preparation, Transition Guelph presented Resilience 2011: A Community Festival, a week-long series of events and workshops culminating on 25 and 26 March 2011 with dignitary addresses from Guelph's mayor and member of Parliament and keynote addresses by *Geography of Hope* author Chris Turner and Wayne Roberts, formerly of the Toronto Food Policy Council. The event included an afternoon of open space planning and other activities and an Earth Hour observance featuring a potluck enhanced by donations of organic food by local farms and businesses, followed by a concert featuring local bands. With its tagline Celebrating Our Community's Achievements, Planning New Ones, the event served the same function as the Great Unleashing described by Hopkins in the deepening pattern (Transition Network 2011); it celebrated the work done by citizens' organizations in Guelph and sought to inspire and mobilize the community to build further resilience. After the success of Resilience 2011, TG volunteers made the Community Resilience Festival an annual event. Festivals in 2012 and 2013, with a greater number of workshops and symposia, skill-building demonstrations and activities, arts and children's activities, and internationally known key-note speakers, were received positively in the community and drew larger numbers of Guelph and area citizens each March, with the Earth Hour pot-luck of Resilience 2013 serving over 250 people.

Run entirely by volunteers and financed largely by the Cooperators Insurance Group and other local donor organizations, the Resilience Festival has a highly collaborative outlook, aligned with a purpose that TG has adopted of supporting and promoting the work of local organizations, thereby culti-vating community resilience. Festival planning includes open meetings and additional conversations with members of other citizen groups, resulting in many synergies, co-promotion, and resource pooling to enhance the visibility and character of the event. Collaboration has included lead-up and follow-up events, the scheduling of festival events to coincide with Sustainability Week and No Impact Week on the University of Guelph campus, Canada Water Week in 2013, and observance of Earth Hour each year. Additional collabora-tions include the Guelph Environmental Leadership's annual Eco-Market as part of the festival, co-promotion of World Water Day with Wellington Water Watchers, partnership with the Alternative Agricultures class at the Univer-sity of Guelph to organize the food logistics of the festival, and generous dona-tions of goods and funds by local businesses. As a result, TG has seen a notable increase in the number of people directly involved as volunteers for the event, of partners (see the sponsors page on the event website at http://www.guel-phresiliencefestival.ca for a complete list), and of core members of the event planning group. The festival has been deemed a success by the organization, with upward of 500 people participating in the Saturday events.

SNAPSHOT IN 2010: INTERVIEW FINDINGS

At the time of conducting TG interviews in 2010–11, TG saw itself and was seen by its partners as not so much a program or an organization as a movement. Three interviewees pointed out that they are not the "doers" per se but that TG is a catalyst that initiates and facilitates action by groups and individuals wanting to contribute to increasing community resilience. Building the resilience of hu-man beings themselves is also seen as important, as is making people-to-people connections. The grassroots element of the initiative was emphasized by many interviewees, one of whom described the self-organizing and bottom-up model as a new way for people to work together. The awareness-raising function of TG was described by one partner organization as "helping people to understand the magnitude of the issue[s] and doing public education, since there is a void of political responsibility," and as "bringing all ages together." Another partner organization noted that TG had put together "lots of workshops and meetings [that were] really informative, very useful, and important."

Figure 6.1. Community potluck in Guelph

Figure 6.2. Community agriculture class

Figure 6.3. Inner Transition workshop

The themes of togetherness and self-reliance permeated the interviews with regard to the vision and objectives of Transition Guelph. When asked to describe the purpose of TG, interviewees from partner organizations responded as follows: "to increase resilience, get people to exchange information, inspire them. Since it's good to have a flexible space to voice concerns, it is a good sounding board for ideas"; "to help prepare people [to] be more self-sufficient and work on mitigation"; "to instigate groups and people that want to start something and support their initiatives under the collective banner of TG." In addition to these points, TG members spoke of resilience, and many mentioned relocalization as part of the purpose of TG. One founding member summed it up as follows: "The purpose of TG is to be a catalyst, to help the community, both individuals and groups, evolve into a low-resource-use, high-satisfaction kind of lifestyle." In fact, one TG participant described the role of TG as mainly community building since a "strengthened, more connected, and thoroughly networked community is what is required to adapt to climate disruption, resource depletion, and economic problems already present that will be worsened for many people."

The most significant changes effected by the organization mentioned by TG members included becoming known by other networks and supporting them, providing a sense of community and go-to place to discuss salient issues, having developed a garden-sharing program, and eliciting positive references to a resilient lifestyle in the Guelph media through radio interviews, guest editorials, and letters by TG members.

A representative of one partner organization stated that the biggest achievement of TG has been to demonstrate to other organizations that it is not duplicating their work, trying to replace them, or competing with them. That claim was received positively by all partner organizations. Another commended them on keeping up momentum and raising interest in the organization.

Recommendations to similar organizations included emphasizing collaboration and cooperation in partnerships and staying positive while learning to go beyond the pain and frustration of dealing with issues such as climate change and peak oil. Collective action was mentioned by one partner organization as key in moving forward: "Transition is very different than other groups that lobby for household-level change. It's like starting to work with your neighbours versus changing a lightbulb."

One participant stated the importance of "creat[ing] a movement where people can come to find allies that believe in a positive sustainable future, not negative messaging creating despair. I see the value of connecting around positive things and having a solutions-oriented and empowering movement."

Some challenges mentioned by TG members, who consistently mentioned the work of the group as a "work in progress" and referred to TG as a learning organization, include the complexity and sometimes denial of issues by a large portion of the public; volunteer retention and the inconsistency of feedback from working groups (with volunteer management identified by many as a weakness of TG and thus a current priority); the unsatisfactory nature of documentation and communication, which limits the potential for tracking and measuring the group's impacts; and, finally, a high reliance on a few key individuals in the first two years, which, coupled with a limited budget, contributed to a limited capacity to face problems of a much larger scale. The formalization of partnerships was seen as one potential solution to limited capacity. On that point, one interviewee said that "TG people aren't the movers and shakers like the people of Eden Mills Going Carbon Neutral; they are the regular citizenry. [TG] is slowly growing and rippling outwards, which is a very healthy way to effect change but happens slowly."

Finally, some lessons learned by the group include the importance of participatory facilitation methods in meetings and of fostering a sense of empowerment. Reinforcing that the movement belongs to all participants and not trying to reinvent the wheel were also mentioned.

COMMUNITY RESILIENCE ANALYSIS

Bridging Social Capital Development – Transition Guelph has adopted the mandate of supporting local organizations working toward greater sustainability and community resilience. TG works closely with a number of other organizations whose members describe their working relationships as positive. Volunteers in TG state that they have learned a lot from Eden Mills Going Carbon Neutral, especially in TG's beginning stages. TG is a member of the Guelph-Wellington Local Food Round Table (it seeks to minimize duplication between the two organizations), the Guelph-Wellington Coalition for Social Justice (it gives input, feedback, and moral and financial support to events such as the Climate Fair in October 2009), and 10 Carden (which provides affordable meeting space and acts as a hub for interactions among social innovation groups). Building Common Ground, Sumac Community Worker Cooperative, the local chapter of Council of Canadians, and other organizations have co-hosted and promoted speakers, films, and informational displays.

TG has an ongoing relationship with the University of Guelph, which hosted the Our Environmental Future conference and provided venues to screen the *In Transition* film as well as Training for Transition. The School of Environmental Design and Rural Development at the University of Guelph has invited presentations and participation in panel discussions by TG spokespersons. The Institute for Community Engaged Scholarship has assisted TG with knowledge transfer, research, and effective practices in event planning for the Resilience Festival and adapting curriculum for the Transition Streets neighbourhood program launched in 2013. TG also collaborated with University of Guelph students on organizing the Life after the Peak conference, a one-day event that gathered eighty high school students from Guelph to learn about and plan for peak oil for the 2010 World Town Planning Day. The Resilience Festival and the university's student-run Sustainability Week coincide in dates and have collaborated each year, co-hosting guest speakers, promoting each event, and providing display opportunities.

TG also has a collaborative relationship (manifested by partners providing venues for events or cross-publicizing them as well as by cross-participation of members in both groups) with the City of Guelph; St. James Apostle Anglican, St. George's Anglican, Dublin Street United, Harcourt United, St. Mathias, and Unitarian Universalist Churches; the Ontario Public Interest Research Group (OPIRG); Guelph Environmental Leadership; and Backyard Bounty. TG was asked to adopt the Appleseed Collective (a local fruit-harvesting and food security group) as one of its practical projects. Another new food-related project is the Treemobile, a fruit tree-ordering group that provides planting by volunteer teams, launched at Resilience 2011. TG is also exchanging ideas and collaborating with other Transition Towns in Canada and around the world.

Other roles for TG that relate to collaboration include identifying partners and gaps so that TG can contribute effectively without "reinventing the wheel" and be a cross-pollinator/connector/facilitator. Relatively few people directly mentioned "undertaking projects" as a role for TG. "Public events," "get the word out," "connect," "inform," "encourage," "be a kind of bridge between groups in the community doing things in sustainability and all kinds of areas" were mentioned more often.

Human Capital Development – In its first two years of existence, TG focused largely on raising awareness and organized many film screenings and speakers events, including an Energy Fair co-sponsored in 2009 with St. James Church's Green Team and a conference on peak oil called Our Environmental Future co-sponsored with the Council of Canadians and city councillor Maggie Laidlaw. It facilitated reskilling workshops (teaching practical skills for relocalization and self-reliance). More recently, skill building has been emphasized further through the work of the Urban Food Working Group and events co-sponsored with several skills-building and "do-it-yourself" groups and Canadian Organic Growers.

TG's education working group organized the Life after the Peak conference and collaboration with the Community Environmental Leadership Program (CELP), an environmental high school program, to organize an afternoon event on climate change and peak oil planning for high school students during the Resilience Festival, demonstrating TG's success in involving youth and students.

Young adults usually make up over 25 percent of the people at general and planning meetings. Resilience 2011 included a day-long EcoAwareness event for the city's secondary school students in which they constructed an oil memorial installation, hunted for "green" products at downtown retailers, learned about volunteer opportunities, and viewed a play produced for the event.

TG's steering group and general volunteers view TG as a learning organization, which means that they value distributing leadership widely and supporting skill development and a sense of empowerment among members, with participatory facilitation in meetings, project initiation by many individuals and small groups, and support of the wider membership for projects under way. Training in governance, communication and decision-making methods, working with diverse communities, and team building in working groups all develop volunteers' capacity to work together successfully. Volunteer management and retention, identified by many as weaknesses of TG, are current priorities and are seen as reducing the high reliance on a few key individuals during the first two years of TG's existence and reinforcing the understanding that the movement belongs to everyone.

Linking Social Capital Development – Guelph City Council was referred to by all TG members as a collaborator. In fact, the Community Development and Environmental Services Departments hosted a TG presentation to inform councillors about the group's goals and start to think about potential collaboration and connection with parts of the municipal structure (especially the Community Energy Initiative). Two councillors, June Hoffman and Lise Burcher, started attending meetings early on and invited steering group members to two sessions on brainstorming and strategic partnerships to take the movement further and to communicate with the city. City councillors are still active in Transition Guelph and have gone to planning sessions for the Resilience Festival. In addition, TG participants have attended the bimonthly meetings of the multisectoral Guelph Mayor's Task Force on Sustainability since its inauguration in the spring of 2010.

The manager of the Community Energy Initiative (CEI) at the City of Guelph saw the role of citizens' organizations as audiences and partners in implementing the CEI. In his mind, community organizations should "move towards delivering services and products" (Kerr, personal communication, 12 January 2011). He gave the example of Guelph Environmental Leadership, an organization that "figured out deliverables by working with the utility [Guelph Hydro]; they have agents delivering green technology products and doing interviewing" (ibid.). When thinking of the future relationship between the CEI and TG, Kerr saw TG as an organization with the role of coordination, which builds synergies among different groups. He mentioned having spoken to TG members on many occasions in the past but having yet to agree on a common tangible project to work on together.

Figure 6.4. Treemobile participants

Figure 6.5. Earthkeepers build outdoor classroom

Figure 6.6. Community outdoor classroom

Figure 6.7. Transition Guelph potluck at an urban garden

EVOLVING LEARNING

Reflections from 2011–2013 – The months since collection of this research have seen TG evolve as an influence on social change in Guelph. Emphasis has been on planning and implementing structures for sustainability of the movement. A volunteer board of directors was formed and conducted TG's incorporation as a non-profit organization in January 2012, and it continues to oversee financial and funding matters. Successful grant applications for Resilience 2012 and 2013 were made, and the creation of two temporary coordinating positions was secured through a provincially funded jobs program; these positions have increased capacity for effective events and organization.

Attention to volunteer recruitment and retention is resulting in easier avenues for engagement. And meetings of volunteers function as social spaces for sharing experiences and interests with people who hold similar views or might be marginalized in the dominant culture. Outreach to new Canadian, youth, Aboriginal, and other population groups, as well as suburban areas of the city, is intended to increase diversity and cross-demographic connections among TG volunteers, bringing the movement for community resilience to a broader spectrum of Guelph residents. Intragroup communication and community promotion are being improved with more effective platforms that make greater volunteer input possible; for example, the TG electronic newsletter is now produced with reports, working group updates, event announcements, and opinion pieces by members.

A second area of emphasis since Resilience 2011 has been on projects that can serve as practical manifestations of the less energy intensive, more satisfying society that TG volunteers seek to create. The Resilience Community Festival has become a successful annual event, with more learning and networking events, greater numbers of volunteers participating in planning and delivery, and more Guelph residents attending. Resilience 2013 included many more skill-sharing workshops and raised interest in continued skills training. The current focus on the local economy began with the launch of a time bank and an associated networking system for offers and requests for services and learning among community and NGO members. New event and training co-sponsorships with a worker cooperative and a public interest research group, as well as ad hoc coalitions, make available a broader range of initiatives for economic literacy and support for local activities. Local food system sustainability is an active area, with the most successful Treemobile planting in 2013, and community orchard projects, workshops on and networks for backyard poultry keeping and food-related skills upgrading, and permaculture research and education are ongoing. TG volunteers and collaborative partners at the University of Guelph, energy utilities, and environmental NGOs have begun preparing for delivery of the Transition Streets program, an "ultralocal resilience initiative" that will create contexts for neighbours on individual blocks or in residential buildings to get acquainted, learn together about living more sustainably, and implement money-saving conservation and efficiency projects in their homes.

Evaluating Progress – Since collection of the research data, Transition Guelph members have continued to observe, invite feedback on, and assess progress of the movement. Collection of demographic and other information has improved somewhat but still remains insufficient for a full picture of which Guelph residents are being reached by programming. TG offers residents support in self-evaluating the resilience of their households via a web-based evaluation tool (http://www.transitionguelph.org/HRA.php), a survey offered to event participants, and exercises in the Transition Streets curriculum. The collection of demographic and other information is being improved with online surveying to better address the needs and interests of volunteers. Visioning and strategic planning events have been held each year. Training for transition is offered periodically, and volunteer meetings offer visioning and "backcasting" activities to apprise new participants of the collective vision for a resilient city and to incorporate their ideas.

SUMMARY OF FINDINGS – TRANSITION GUELPH

DEVELOPMENT OF BRIDGING SOCIAL CAPITAL	Healthy working relationships with other organizations (Council of Canadians, Wellington Water Watchers, Guelph Environmental Leadership, 10 Carden, and University of Guelph)
DEVELOPMENT OF LINKING SOCIAL CAPITAL	Participation in the Mayor's Task Force on Community Energy Greetings at Resilience Community Festival opening by Guelph mayor Dialogue with Community Energy Initiative manager
DEVELOPMENT OF HUMAN CAPITAL	Awareness-raising events on peak oil and climate change through film screenings, speakers events, and conferences Youth education activities, Life after the Peak conference, collaboration with Community Environmental Leadership Program Numerous educational resources posted on website and in newsletters

Table 6.1. Summary of findings

SWOT ANALYSIS OF TRANSITION GUELPH

STRENGTHS	A very participatory process in planning for the Resilience Festival shows a commitment to collective action and engagement of community resources.
	TG is appreciated by members for its social support and networking function, which builds bonding social capital.
	TG increasingly has been taking on the "hub" function and building bridging social capital by convening three joint meetings of various NGOs in Guelph, which also stimulated collective action on planning for the Resilience Festival.
	TG has granted an important place for cultural capital in events by integrating food, dance, music, bicycling, theatre, and public art.
	Recent improvements in decentralizing planning by adding new members to the steering group, inviting more people to meetings, and using technology for event planning and outreach (Google sites, social networking) create more truly collective action rather than having a small group of individuals doing most of the work.
	Representation and participation on various bodies, such as the Mayor's Task Force on Community Energy and the Guelph-Wellington Coalition for Social Justice, increase the group's visibility and allow active agents to exchange their visions.
WEAKNESSES	To date, there has been too much focus on building capacity to act (community resources) and not enough action taken (resource development and engagement).
	In the first two years of TG's existence, weaknesses in communication, lack of a clear decision-making process, inadequate follow-up with current and potential volunteers, and inconsistent meeting management jeopardized the engagement of human capital.
	Organizational weaknesses mean that strategic action processes could be improved to include project management systems, and priorities could be operationalized into organized plans of action.
OPPORTUNITIES	There are opportunities to build political capital by planning projects related to the CEI at the Resilience Festival.
	Bridging social capital can be developed with less environmentally minded groups and people through the new Engagement Committee and its "ambassadors" and the use of a community-based social marketing approach to program delivery.
	TG can foster new working groups and projects after the Resilience Festival to ensure that collective action continues.
	The Open Space session at the Resilience Festival can be used to quick-start the process of the Energy Descent Action Plan, a great tool for strategic and collective action.
THREATS	The use of certain transition-specific terms, such as "great unleashing," is not conducive to outreach among the "mainstream" and thereby building bridging social capital; as a result, the group might remain in a niche.
	Perception as a facilitator (as opposed to a leader) of community resilience by TG members and Guelph City Hall might limit the scope of future partnerships between these active agents.
	A focus on capacity building slows down the process of developing linking social capital.

Table 6.2. SWOT analysis of Transition Guelph

CONCLUSION

The research presented in this chapter sought to answer the question *how are community organizations seeking to build community resilience to climate change and peak oil in Ontario?* Findings are summarized in Table 6.1.

Given the short period of time in which TG has been in operation, it is too early to provide a definitive evaluation of the impact that the organization has had on community resilience. Based on the empirical data from this research and informed by the literature on community resilience, especially the eight dimensions of the Magis framework presented in the previous chapter on Eden Mills Going Carbon Neutral, we summarize our impressions of the synthesis of internal strengths and weaknesses, and indicate some external opportunities and threats in the future for TG, in the form of a SWOT analysis in Table 6.2. This table was used in a workshop session at a Community Research Report Back event held at the University of Guelph in April 2011 to guide a dialogue on concrete actions for "what's next?" It was also presented as a working tool for the group, understanding that community resilience building is a learning process that unfolds over a long period of time.

The research on Transition Guelph in this chapter indicates that there are reasons to be hopeful when speaking of citizen-driven community resilience, especially since the community-led responses to increasing resilience to climate change and peak oil analyzed here are in their early days. Much can be learned from cases of citizen organizations already engaged in responding to the issues by increasing community resilience. Organizations such as Eden Mills Going Carbon Neutral and Transition Guelph show great potential and are doing important work. In the face of the predicted impacts of climate change and peak oil, every community and municipality in Ontario ought to consider its own responses to build community resilience.

References

Colussi, M., and P. Rowcliff. 2000. *The Community Resilience Manual: A Resource for Rural Recovery and Renewal.* Vancouver: Canadian Centre for Community Renewal.

Haxeltine, A., and S. Seyfang. 2009. "Transitions for the People: Theory and Practice of 'Transition' and 'Resilience' in the UK's Transition Movement." Tyndall Centre for Climate Change Research, Working Paper 134.

Hopkins, R. 2008. *The Transition Handbook: From Oil Dependency to Local Resilience.* Totnes, UK: Green Books.

———. 2011. *Rethinking Transition as a Pattern Language: An Introduction.* http://transition-culture.org/2010/06/04/rethinking-transition-as-a-pattern-language-an-introduction/.

Transition Network. 2011. *Patterns/Transition Network.* http://transitionnetwork.org/patterns.

CHAPTER SEVEN

THE CHANGING FACE OF AGRICULTURE: CASE STUDIES OF ORGANIZATIONS BUILDING A NEW FUTURE FOR FOOD

ERICA FERGUSON

INTRODUCTION

Global forces and trends in climate and energy are increasingly shaping all communities, including those in rural southern Ontario. As these forces shift, so must our communities, and community organizations must find strategies to respond and adapt to changing circumstances. In general, response and adaptation work is seen as resilience building, which certain innovative organizations are more proactively pursuing at the local level. This chapter highlights two selected case studies of resilience: Everdale Environmental Learning Centre and FarmStart. Each has a unique approach to adapting agriculture to reflect new realities. These two case studies explore community resilience responses as they relate to small farms and training for resilience.

The need to increase the resilience of the food systems that we most rely on for basic sustenance is central to increasing community resilience. From the vibrant "Eat Local" campaigns, including farmers' markets, community-supported agriculture, and hundred-mile markets, to increasing students' understanding of food issues, there are a number of entry points into resilience planning through a food system lens. Community organizations exploring farming issues and food production comprise a key component of resilience planning.

CASE STUDY I: EVERDALE

In Erin Township, just south of Hillsburgh, is a fifty-acre organic farm known as Everdale Environmental Learning Centre. In a bucolic setting of rolling hills nestles a charitable centre dedicated to sustainable farming and education. The executive director, Brendan Johnson,[1] describes Everdale as a farm learning centre that trains the next generation of farmers and educates youth to be stewards of the land, environmentally aware local food champions, consumers, and future farmers.

History – Since 1998, Everdale has been providing hands-on, practical learning experiences to move toward environmentally focused agricultural practices. It evolved over the years with the needs and wants of its constituency and found gaps in educational opportunities both for new farmers and for children to have experiential farm opportunities. The organization of Everdale was created around those needs. Currently, Everdale predominantly serves the local constituency of the Wellington County area, with approximately 70–75 percent of visitors from the local area. Other participants come from the Greater Toronto Area, and approximately 5 percent are provincial or national visitors. Everdale, as a place and an organization, is specifically designed to provide educational experiences for people to learn about farm and food issues; it is committed to connecting consumers and farmers.

The successful Community Supported Agriculture (CSA) program began in 1998 with thirteen members; it currently has over 300 members. CSA programs offer individuals the opportunity to purchase a "share" of the vegetable harvest at the beginning of the growing season, which provides each shareholder with a weekly supply of fresh produce. With incorporation as a not-for-profit organization in 2000, Everdale began to offer farm internships, which were closely followed by farm programs for children and youth. Everdale was a founding partner of the Collaborative Regional Alliance for Farmer Training (CRAFT) in 2003 and received charitable status in 2004.

As an educational facility, Everdale has a number of projects, including twelve acres of organic vegetable production, laying hens, sheep, donkeys, draft horses, compost piles, solar showers, greenhouses, an earthen oven, a wind turbine, and solar dehydrators. There are four straw-bale buildings, a learning space called "the Hub," "Home Alive" (a demonstration naturally built home), and two staff residences.

Everdale Programs

Internships and Farmers Growing Farmers – Everdale runs a successful internship program that teaches organic farming methods in the most practical setting—on the farm. After running the internship program for some time and seeing many aspiring farmers start their own farms, only to find itself without the business plan that it needed for success, Everdale started the Farmers Growing Farmers (FGF) program. FGF works with new farmers who are pursuing farm enterprises. It helps with planning, mentorship, start-up, and establishment of ecologically based farms interested in direct marketing. Eventually, Everdale aspires to get formal recognition for new farmers and the training that they receive through Everdale as well as access to land, grants, and loans. At this point, for many new farmers, farming is cost prohibitive. Everdale's promotion of new farmers and the success of new farmers in their communities are showcasing ecological farming as a real way to grow food and

ure 7.1. Everdale Environmental Learning Centre

gure 7.2. Harvest time at Everdale Environmental Learning Centre

feed communities. Everdale shows how it can be successful to farm without the inputs of conventional farming.

School Programs and Summer Camp – Through workshops that are hands on and linked to the public school curriculum, Everdale works to develop environmental stewardship. It both teaches schoolchildren on site and visits schools to provide interactive environmental programming. Beginning in late fall and running until early spring, Everdale runs the "Farmers in the Schools" programs. Everdale educators (and sometimes an Everdale chicken) bring farm learning into the classroom with curriculum-linked programs. During the summer, Everdale offers a farm camp that has been very successful.

Workshops – A number of skill-building workshops are offered throughout the season. They include a series of programs, among others, on backyard chickens, beekeeping, food preservation, growing sprouts, permaculture, invasive species, garden maintenance, bread baking, mushroom cultivation, soap making, onion braiding, season extension, and seed saving.

Dedication to environmental sustainability, community development, and resilience building has positioned Everdale to be recognized for its work. In 2011 alone, it was awarded the Sustainable Farm Award from the National Farmers Union and the Education Award from the Organic Council of Ontario. In the same year, the chair of its board was short-listed for the Local Food Category of the Green Toronto Awards.

CASE STUDY 2: FARMSTART

FarmStart is on a mission: to support a new generation of farmers "to develop locally based, ecologically sound and economically viable agricultural enterprises" (FarmStart 2011). FarmStart actively searches for opportunities to increase local food supply by supporting agricultural entrepreneurship and new farm enterprises. FarmStart provides training on sustainable business models that allow new farmers to have sustainable livelihoods. Its website is clear that food security is dependent on recruiting and training new farmers to farm successfully: "The loss of farmers and the lack of young people taking their place will soon become a very real problem for the Canadian domestic food supply. With an average age of farm operators at 52, and some 50% of current farmers looking to sell or transfer their farms in the next ten years, agriculture faces difficult succession / intergeneration transfer issues. The difficulties, risks and disincentives facing those who wish to start a farm enterprise are often overwhelming and discouraging" (FarmStart 2011).

There are four groups that FarmStart works with: young people with farm backgrounds, young people new to farming, second-career farmers, and new Canadian farmers. The types of farm enterprises supported vary from urban to rural as well as cooperative farms, intensive agriculture, no-till organic agriculture, and those that explore new market opportunities or value-added products. The people whom FarmStart supports tend to be evenly split among young people (mostly from non-farm backgrounds), second-career farmers, and new Canadians.

Executive Director Christie Young (2011) describes the huge transition occurring in agriculture, with 50 percent of farms being transferred in the next ten years, 75 percent of which do not have successors. There is a generational impasse that Young sees as dovetailing with a transition in the approach to farming, which is linked to fossil fuels, health, quality of food, and equity. The new generation of farmers is drawn to a new kind of farming that binds the entrepreneurial to the ecological. Young finds the nexus of world issues in food: "If we can't figure out how to feed ourselves, to grow food and treat the people who grow food properly, and treat the land properly, we won't be able to figure out all the other issues." Young sees food and agriculture as central and FarmStart's role as investing in new farmers to increase community resilience in food.

FarmStart promotes ecological agriculture and sees the role of the farmer as a steward of the environment. This encompasses all aspects of the farm enterprise and sits in contrast to conventional agriculture: "In a time of rising oil costs and decreasing fossil fuel supplies industrial farming is becoming less environmentally and economically sustainable" (FarmStart 2011). In general, the farming style promoted by FarmStart hinges on diversity and management that builds the soil through the use of cover crops, green manure, compost, mulch, crop rotation, and no-till methods. FarmStart advocates for organic methods and supports certified organic agriculture, yet it does not exclude other approaches.

History – FarmStart began in 2005 just north of Guelph, Ontario, at the Ignatius Jesuit Centre, which owns hundreds of acres and was looking to do something different. Young, the founder of FarmStart and the current executive director, had visited Intervale (see Intervale 2011) in Burlington, Vermont. Intervale was one of the first incubator farms that she saw in action (it has since evolved into a community farm). What she saw was a physical facility that supported farmers, with a community and an intention that were impressive: it was working to help new farmers learn about farming and establish themselves. When Young returned, many of her friends were starting farms, and she witnessed little support for them in their transition. There were internship programs, and the local food movement was thriving, but, in terms of support to make a farm business a reality (support for new farmers with a new business model), there was nothing.

The Ignatius Jesuit Centre was the site of the first FarmStart incubator, and FarmStart offices are still located there. FarmStart quickly realized that a farm incubator was not enough, for people came to the incubator without the business skills required for a farm enterprise. Thus, FarmStart developed programs in response to the needs of the new farmers to help them develop businesses that were sustainable economically as well as ecologically and that were modelled for increasing resilience in terms of inputs or fossil fuels.

FarmStart Programs – FarmStart provides training and resources as well as more tangible farm start-up resources, including access to land, farm infrastructure, and a small grants program. With the grants come business plan development and review. One of the strengths of its work is the mentorship model that it promotes through its programs, in which new farmers are connected to established farmers to learn, ask questions, and start to build a farmer network. This mentorship connection is described by Young as another form of capital—social and human capital that enables new farmers to increase their confidence and to build a community of support for their work.

FarmStart also works at farm linking to increase access to land for new farmers and to address some of the issues of succession facing farming communities in Ontario. This is one of the most challenging aspects of FarmStart's work—how to transfer from a farming community that uses traditional, fossil fuel–based methods to a farming community that is quite different. FarmStart views its role as that of a bridge builder—not to take on the issue of succession entirely but to create a different way of looking at the transition, to provide tools, resources, and encouragement, and to help form relationships of trust. The area of farm linking is key, from working with farmers who believe that something different is possible, to assisting with a relationship between farmers who have land and new farmers, to transferring land. There are many challenges to farm linking, not least of which are large farm assets (equipment, machinery) that might not be relevant to the next generation of farmers with less reliance on fossil fuels.

The new farmers engaging with FarmStart feed into the increasing relevance and popularity of local food combined with concerns about animal welfare, ecological sustainability, climate change, and peak oil. Provincially, all of these factors have resulted in a remarkable increase in the number of farmers' markets and in the number of farmers offering CSA as an option for fresh produce purchases. In the Greater Toronto Area, the McVean Incubator Farm has been making waves as an incubator of new farmers. It is located in Brampton on conservation land and was formed through an agreement between Farm-Start and the Toronto and Region Conservation Authority. At McVean, people interested in pursuing a career in agriculture have access to test plots and are

THE GENERAL OBSERVATIONS GAINED FROM BOTH EVERDALE AND
FARMSTART AS FARM SUPPORT ORGANIZATIONS ARE REPRESENTED BELOW:

GENERAL OBSERVATIONS	EVERDALE CASE STUDY	FARMSTART CASE STUDY
Recognition of climate change and the reliance on fossil fuels for agricultural production is a driving force for small-farm support.

Recognition of small-farmer support work is growing, even though financial support continues to lag behind.

A localized approach to farmer education and local food networking is expanding, based to some degree on knowledge of climate change and energy needs.

Local businesses and local jobs come from supporting small-farm development, which relies on a number of different approaches.

Ecological, small-scale farming is a necessary transition in agriculture. | Climate change and peak oil are deeply interwoven into the approach to programs at Everdale. It specifically provides programs to increase people's ability to grow food to have healthier communities and people. A driving force behind its work is the understanding that current farming methods are not tenable in the future—that a different way of farming is needed.

A major challenge for Everdale, as a charitable non-profit organization, is funding. Many people contact Everdale to learn about alternative farming methods. Finding the resources to continue its educational programs in the non-profit environment is extremely challenging.

With a long-term focus, Everdale is looking to influence other organizations, both provincial and national; its vision is to assist in training other organizations to provide farmer training and youth programming to create their own food hubs and networks. Everdale has a vision of a network of local food farm communities spreading across the country and across the world.

Everdale sees its work supporting more local, sustainable, and ecological farms as being directly related to creating more local businesses and jobs as well as building community. As Johnson (2011) elaborates, "the act of growing our own food to feed our own people leads directly to healthier communities as people gather to grow food and to support local farm businesses. This in turn increases communication, and people get to know each other outside the direct link to food."

The type of farming that Everdale teaches and advocates is also directly linked to the challenge that peak oil poses to conventional farming methods. Locally produced food— food grown within a community for that community—lessens the dependence on food that needs to be shipped from other countries. | FarmStart's work hinges on acknowledging that the current system of agriculture is deeply embedded in fossil fuel reliance and methods that are not ecologically sound. FarmStart believes that the consolidation of food business has worked against a resilient food system, putting farmers at risk as they rely on fewer types of crops, putting ecosystems at risk through monocultures, and putting food safety at risk through consolidation of production lines.

FarmStart has put in place necessary partnerships, such as working with the Toronto and Region Conservation Authority, to gain access to land for a small-farm start-up plot, and it has achieved charitable status. The support for FarmStart's work continues to grow, but the focus on partnership and funding is a continual requirement.

Currently, the agricultural system is extremely reliant on cheap energy since the amount of energy put into food production greatly outweighs the amount of energy in the food. FarmStart recognizes that farmers more than most people feel the effects of rising fuel costs because their businesses are so reliant on energy. The resilience and sustainability of farms increase as they are less reliant on fossil fuels. FarmStart provides alternative ways of growing that are less dependent on fossil fuels.

As FarmStart aims to diversify food production, it trains people in a new approach to growing, one that is increasingly diverse: a type of farming that promotes ecological as well as economic resilience. FarmStart envisions a farming industry that has many people growing food, with many options for how it is grown and processed. The links among food, public health, ecological health, and agriculture shape the work at FarmStart.

Young (2011) believes that farmers will feel the effects of climate change most directly as weather patterns influence farming more than other occupations. She suggests looking at business models that change the predominant type of farming—models that increase the resilience of ecological systems and cut yields in the short term but are more sustainable in the long term. |

Table 7.1. Farm Support Organization Observations

connected to business training and agricultural support. Many new Canadians, with farm experience in their countries of origin, have used McVean to explore their possible participation in agriculture in Ontario. The interest in farming tends to stem from new Canadians' connection to food as well as many other factors, including the desire to undertake meaningful work and meet the food needs of their communities as well as other markets.

FarmStart does its best to take the rose-tinted glasses off farming. Young (2011) acknowledges that the greatest challenge to its work is public understanding—that farming is really hard work and that "the general population is used to cheap food that fossil fuels and off-shore labour have allowed." It is evident to FarmStart that, no matter how passionate and dedicated a new farmer might be, it is hard to figure out how to make a farming business work in this context. The issue of land ownership also weighs heavily; land values are almost always outside the reach of new farmers. Until the price of food reflects the true cost of producing it—a cost that will inevitably increase as reliance on fossil fuels is increasingly replaced by human or animal labour—farmers are stuck growing food for a clientele used to paying low prices.

CONCLUSION

In times of uncertainty, innovation takes on increasing importance. As the farming sector continues to face challenges and changes, the need for effective responses increases. In the cases of Everdale and FarmStart, farmer education and encouragement are beginning to address the need for more farmers and more sustainable farms that are resilient to changing climate and energy regimes. As such, Everdale and FarmStart provide case studies of innovation at the not-for-profit level that is spawning increased innovation on the ground in the agricultural sector. These initiatives create possible pathways to sustainable food systems and resilient rural communities.

Resilience also happens at the individual level, and the next chapter provides the perspective of one farmer who has attempted in his own way to respond to the challenges of climate change and peak oil over the forty years of his career.

Notes

1   At the time of writing, Johnson was the executive director. He is no longer with Everdale, but all inquiries about this case study can be directed to Gavin Dandy, Everdale farm director and founder.

References

Everdale Organic Farm and Environmental Learning Centre. 2011. http://www.everdale.org.

FarmStart: Supporting a New Generation of Farmers. 2011. http://www.farmstart.ca.

Intervale Centre: Sustaining Land, Sustaining People, Sustaining Farms. 2011. http://www.intervale.org.

Johnson, B., Everdale executive director. 2011. Interview with the author, 1 September.

Young, C., FarmStart executive director. 2011. Interview with the author, 26 August.

CHAPTER EIGHT

ONE FARMER'S APPROACH TO BUILDING RESILIENCE

TONY MCQUAIL

Farming is the practice of turning sunlight into substance. Through our interaction with the ecosystem, we harness the power of photosynthesis to produce food, fibre, and fuel. Although we take this miracle for granted in North America, it is the essential activity for any society. If you don't have food, then you don't have a society for very long. If you don't figure out how to raise that food without degrading or destroying the ecosystem on which agricultural productivity depends, then you leave crumbling monuments and eroded hillsides. Until relatively recently, farming was a solar-powered activity, but, in the period since the end of the Second World War, it has become a reprocessing activity, converting petroleum into food.

Many of our current problems stem from our failure to understand and accept that we are biological organisms on a finite planet. We experienced a brief moment in history when we seemed to be able to step outside those constraints, and this experience has coloured our assumptions of what is real and what is normal. In one century, we have burned through millions of years of accumulated biomass in the form of fossil fuels. Our beliefs in economic growth and mechanical progress rest on this conflagration. It seems obvious to me that we cannot sustain these levels of energy use with annually renewable sources. But what seems obvious to me seems unthinkable in most of the discussions of how to address climate change, peak oil, and environmental degradation. If we are going to rebound rather than crash from these challenges, then we will need to develop an agriculture that *restores* our ecosystem and our rural communities while using vastly less energy.

In the early 1970s, I was farming and in the environmental studies program at the University of Waterloo. I looked at the research on energy productivity of different systems. The energy return on energy invested (EROEI) or net energy productivity is the ratio of energy that comes out of a system divided by the energy put into it. It was fascinating to compare pre-industrial with industrial agriculture and food systems. Pre-industrial systems showed an EROEI of between five and fifty. That is, for every unit of energy put into the system, between five and fifty units came out of it. In pre-industrial agriculture, that energy was human labour, draft animals, tools, and seeds saved from previous crops. The high end of the scale reflected intensively managed and layered systems such as paddy rice. The low end was simple subsistence agriculture. But to me, what was interesting was that agricultural systems did not go lower than five units out per unit in. My guess is that an agricultural system that produced fewer than five units literally "starved out"; it did not yield enough surplus energy to have a reserve for bad harvests or to raise the next generation.

Industrial agriculture, with its fertilizers, herbicides, pesticides, diesel fuel, big machines, and transportation, processing, and distribution networks, has an EROEI of 0.1. *In other words, ten units of energy are used in the system to get one unit of energy to the table.* Industrial agriculture is a system for converting petroleum into food in an extremely wasteful fashion. Unfortunately, industrial agriculture's carbon footprint is even larger than its EROEI, for it has proven to be a system that also converts soil organic matter into atmospheric carbon dioxide.

Tragically, what we have done with industrial agriculture has been echoed across our whole economy, in which we have redesigned our activities to use ever greater amounts of energy as we replace labour with fossil fuels. When we first started this substitution, the EROEI of petroleum was impressive. Early oil wells often produced over 100 units of energy for every unit spent in drilling. It was the easy oil to get to. By the 1970s, the EROEI had declined to thirty units out for each unit in and continues to drop (Heinberg 2003, 138; see also Cleveland et al. 1984; Gever et al. 1991; Odum 1996). The tar sands in western Canada might be getting down to one out for one in if you count all the hidden subsidies. As the EROEI decreases, environmental impact increases, and the driver of our past 100 years of economic growth and our ability to fuel industrial agriculture collapse. Without a high EROEI, it is impossible to achieve the rate of growth that we assume denotes a healthy economy. Trying to achieve those rates of growth with low EROEI energy systems will be irreparably destructive and counterproductive.

We are not going to create a sustainable society by feeding our food to our machines. We are going to destabilize society completely if we plan to take the food out of the mouths of the poor to put it into the tanks of SUVs and jet planes. We will also continue to destabilize the ecological life support systems of this planet. But we are reaching the point of "peak oil" or, as Richard Heinberg (2007) has written, "peak everything." What can we do?

The answer seems to me to be right under our noses. We need to redesign our economies, societies, and food system to run on the energy that goes into our mouths. And we need to remember how to produce that energy (call it food for ease of comprehension) in a manner that yields an EROEI of five or more. As a society, we need to develop an ecological agriculture around and within our urban centres in which food is grown with a minimum of energy inputs and a maximum of ecological design. We need to redesign our cities to be walkable, bikeable, breathable, and livable, where most of the energy to make the city function comes from the food that we eat. If we did so, then we could likely use photovoltaics, wind generators, and methane digesters and convert some biomass into liquid fuels to provide the energy to run public transit and communication technologies and even some tractors and combines in larger farm fields. And we could use our remaining petroleum far more carefully to bridge the gap between where we are today and where we need to be if we are to have a tomorrow.

This is what we set out to do in the mid-1970s on our farm. We had grown up with cheap gas, twenty-five cents per gallon, but we had also lived through the US oil peak, the Arab oil embargo, and empty gas pumps. We wanted to design a farm that was more ecologically sustainable and that would run largely on the energy produced on the farm. I have been a farmer for over forty years; I have been interested in renewable energy for all of those years. In the 1970s, we built a passive solar home. We put up the first modern, interconnected wind generator on the Ontario Hydro grid in 1978. We were using photovoltaic panels to run electric fencers more than twenty years ago and currently use them to run our livestock water and garden irrigation in the summer as well as a ten kilowatt microfit array. We formed a co-op with some other farmers and tried to make an ethanol still but were unsuccessful. We bought a team of horses for farm power that could run on homegrown renewable fuel, and we switched to ecological and organic farm practices. As well, we do a lot of the work on our farm with our own muscle power—no need for an exercise machine or fitness centre.

In 1973, we started with a rundown farm that had been cash-cropped for many years. After several years of farming it relatively conventionally, we decided to switch to an organic approach and to buy a team of horses. We've experimented with a variety of renewable energy technologies, and true horsepower is one of the more "off the shelf," ready-to-use varieties. The greenhouse gas, economic, and environmental advantages of a power source that runs on homegrown biofuels, where the CO_2 is in an annual cycle, encourages soilbuilding sod crops in the rotation, produces a rich organic soil amendment, and can be renewed by biologically manufacturing a replacement on the farm are significant.

After taking a holistic management course in 1995, we decided to market our farm's production directly and developed a small meat retail business and a Community Supported Agriculture (CSA) garden. We were able to stop

Figure 8.1. Tony's horses and solar panels

Figure 8.2. Discing with a four-horse hitch

Figure 8.3. Fran McQuail cultivating

Figure 8.4. Using a walking plow

Figure 8.5. Tony cultivating

working off the farm and have shown a profit every year since. We also have a community of customers who appreciate our efforts and encourage us. They enjoy visiting the farm, where they can see perennial crops protecting the soil while feeding the pastured livestock. They can listen to the birds and insects singing in the windbreaks that we planted years ago.

To me, it is hopeful that actions addressing one of the challenges often address all three. We increase the resilience of our land and our farm economy as we transition to ecological agriculture. Increasing soil organic matter improves the land's ability to absorb and retain rain, making our farm more resilient to floods or droughts. It sequesters carbon from the atmosphere in the soil, making a positive contribution to climate change. It reduces our reliance on off-farm inputs, improving our economics. By avoiding toxic pesticides and harsh chemical fertilizers, we improve the health of the soil and the local environment for pollinators, birds, amphibians, and beneficial insects. By having a direct relationship with our customers, we were unhurt by the BSE (bovine spongiform encephalopathy) crisis.

Modern industrial agriculture, with its reliance on fossil fuels, its negative energy productivity, and its conversion of soil organic matter into atmospheric carbon dioxide, is a dead end in an era of peak oil and climate disruption. We need to remember that farming is about converting sunlight into food and then reimagine how to do it using the wisdom of the past as well as the insights of the present. For some resources to help with that imagining, check out Holistic Management International (http://holisticmanagement.org/), whose slogan is "Healthy Land, Healthy People, Healthy Profits," and Mark Sheppard's book *Restoration Agriculture: Real-World Permaculture for Farmers*.

References

Cleveland, C.J., R. Costanza, C.A.S. Hall, and R. Kaufman. 1984. "Energy and the U.S. Economy: A Biophysical Perspective." *Science* 225: 890–97.

Gever, J., R. Kaufman, D. Skole, and C. Vorosmarty. 1991. *Beyond Oil: The Threat to Food and Fuel in the Coming Decades*. Boulder: University Press of Colorado.

Heinberg, R. 2003. *The Party's Over: Oil, War, and the Fate of Industrial Societies*. Gabriola Island, BC: New Society Publishers.

———. 2007. *Peak Everything: Waking Up to the Century of Declines*. Gabriola Island, BC: New Society Publishers.

Odum, H.T. 1996. *Environmental Accounting, Energy, and Decision Making*. New York: John Wiley.

Sheppard, M. 2013. *Restoration Agriculture: Real-World Permaculture for Farmers*. Austin, TX: Acres USA.

CHAPTER NINE

PRACTICAL RESPONSE: OPTIONS FOR AGRICULTURE

MARGARET GRAVES, BILL DEEN, EVAN FRASER, AND RALPH C. MARTIN

WHY ARE THERE CONCERNS ABOUT LACK OF RESILIENCE IN THE AGRICULTURAL SYSTEM?

Among the major challenges facing Canadian agriculture over the next fifty years, there are twelve that may decrease the ability of farmland to produce adequate food, feed, fibre, and fuel in the future (Deen et al. 2013). These challenges include peak oil, reducing greenhouse gases (GHGs) to mitigate climate change, adapting to climate change, the loss of functioning ecosystems, peak phosphorus, water quality, water supply, soil degradation, population growth and demographic change (alongside increasing demand for agricultural products), loss of dependable agricultural land, loss of biodiversity, and food waste.

Some of the challenges listed above stem from the trends that we have seen in agriculture over the past 100 years. In North America and Europe, agriculture has intensified, with substantial increases in total production and productivity (production per unit land, per farm, and per animal; ibid.). For most of the past century, cultivated land[1] area in Canada increased, including a 40 percent increase in cropped land[2] area from 1951 to 2011 (calculated from CANSIM 2013; see Figure 9.1). That trend has reversed, and the area of cropped and cultivated land has decreased since 2001, partly because of the conversion of agricultural land to other uses (Hofmann, Filoso, and Schofield 2005) but also because increased productivity of land use has mitigated any loss of production. Alongside increased productivity have come shorter crop rotations and decreased natural biodiversity through large-scale cultivation of a few staple grain crops (in particular wheat, corn, and soybean). Recently, decreased areas of small grains, forages, and legumes in favour of more intensive crops (canola, corn, and soybean; see Figure 9.1) have been associated with a dependence on nitrogen fertilizer for soil fertility and soil organic matter (SOM) from crop residues, decreasing the soil's ability to act as a buffer to shocks such as pests or weather events. Increased use of nitrogen fertilizer to promote high yields can also result in excess nutrient discharge from cropped fields, negatively affecting the agro-ecosystem (Janzen et al. 2003; Smil, 2000,

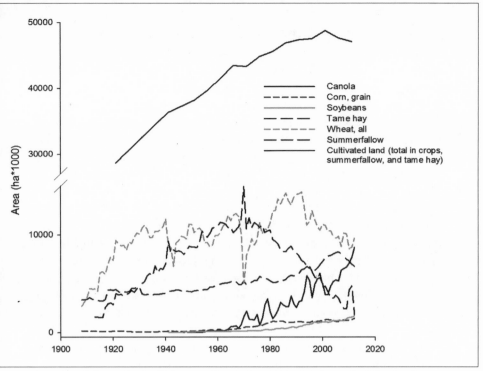

Figure 9.1. Historic seeded area of major field crops in Canada, from 1908–2012. Source: Statistics Canada

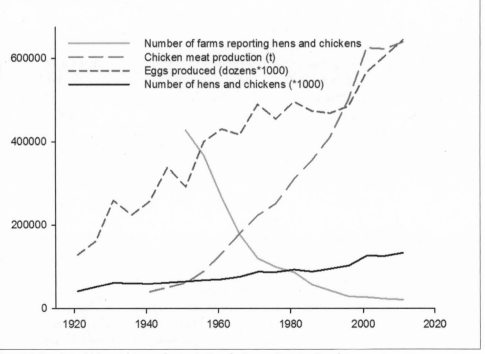

Figure 9.2. Trends in chicken and egg production in Canada. Source: Statistics Canada

2012). Areas that produce corn, wheat, and soybean export regionally to cities, as food, or to concentrated livestock operations, as feed, resulting in large-scale export concentration of nutrients.

Figure 9.2 uses the example of Canadian poultry production to illustrate increases in livestock productivity and concentration of livestock on a small number of farms. Application rates of nitrogen and phosphorous (N and P) might be higher on land close to livestock operations since high transportation costs might preclude shipping manure farther afield (Eilers et al. 2010; Van Horn 2002; Zebarth et al. 1998). Canadian ecosystems such as the Great Lakes–St. Lawrence River and Lake Simcoe–Holland River in Ontario, and the Fraser River Lowland/Abbotsford aquifer in British Columbia, highly populated and intensively farmed, have had significant nutrient-loading issues. Rivers, lakes, and groundwater in these areas have experienced high nitrate levels, algal blooms, eutrophication, and dead zones because of nutrient (N and P) discharge (Chesnaux, Allen, and Graham 2007; Evans et al. 1996; Great Lakes Commission 2012).

Two major challenges stood out as not only urgent but also requiring a radical transformation of the Canadian agricultural system (Deen et al. 2013). The first is the need to ensure that Canadian agriculture helps to build neglected agro-ecosystem health and ecosystem function. The second is that Canadian agriculture needs to become more adaptive to climate change. Although these two challenges were independently selected, they are also connected. The trends in mainstream agriculture in Canada, briefly elucidated above, have degraded the agro-ecosystem and its functional role to provide a stable base for farming. This ongoing degradation and its momentum exacerbate the second challenge—lack of adaptive capacity to climate change—since the resilience of the agricultural system to shocks from climate change depends largely on ecosystem health. In light of the Deen et al. (2013) analysis, this chapter explores these two topics in detail, including practical solutions to address each crucial challenge.

CLIMATE CHANGE ADAPTATION

Although climate change is a nebulous challenge, generally it can be broken down into two categories: mitigation and adaptation. Mitigation involves decreasing GHG emissions or increasing GHG sequestration (pulling GHG out of the atmosphere and storing it) to reduce atmospheric GHG. This is important for mitigation of the greenhouse effect, which contributes to increased climate variability (Smith et al. 2007). However, reducing atmospheric GHG is a slow process, and climate variability is increasing. More frequent weather events such as floods, droughts, and storms caused by climate change pose an immediate threat to our food supply, requiring urgent adaptive action. Certainly, agriculture should strive to be a leader in reducing GHG emissions and sequestering GHG (see ibid. for more information), but adaptation (previously noted to require both transformative and urgent change) must happen faster to minimize impacts on agricultural production.

At the global level, atmospheric CO_2 levels were at 390 ppm in 2010 (Stocker 2010) and are predicted to rise to 550 ppm by 2050 (Prather et al. 2001). Although CO_2 has a stimulatory effect on plant growth (Jaggard, Qi, and Ober 2010), the effect is tempered by a range of other impacts stemming from climate change (see Table SPM.1, IPCC 2007, 18, for projected impacts). The effects of climate change are unpredictable and will differ widely across the globe. Atmospheric ozone levels are also rising and can be detrimental to plant growth (Hayes et al. 2010; Jaggard, Qi, and Ober 2010; Royal Society 2008). Increased temperatures might stimulate yields (longer growing seasons) or hamper them (early maturity). The IPCC (2007) predicts that increases in temperature will have an overall detrimental effect on crop yields in low-latitude developing countries but might stimulate yields in mid- to high-latitude developed countries. Li et al. (2009) estimate that the global risk of drought will double by 2050, and heavy rainfall events seem to be on the rise (Gornall et al. 2010). Inefficiency of input use during drought years is more problematic since lower-yielding crops take up fewer nutrients; this reflects more danger to the ecosystem since residual soil nitrogen is higher in those years (Eilers et al. 2010). With uncertainty and variability of weather events, our agricultural system must be able to adapt and bounce back. Changes in weather and climate will also affect the incidence and severity of insect pests, pathogens, and invasive species (Adamo et al. 2012; Dukes et al. 2009).

Strategies to adapt to climate variability require transformative change away from simple corn-soy rotations so common in North America. More complex systems have been shown to be more resilient (i.e., to maintain productivity in situations of water deficit or excess soil moisture) and less dependent on nitrogen fertilizer to sustain yields (Drury and Tan 1995; Liebman et al. 2013). Varieties resistant to various stresses need to be developed and made available to farmers. Drought resistance is built through undertaking soil stewardship, building SOM, and using hardy varieties, while a resilient agro-ecosystem comprises greater diversity, including set-aside areas, greater integration of farming systems, and expanded crop rotations. These strategies will expand the base on which agriculture rests and should allow for more consistent year-to-year crop yields.

AGRO-ECOSYSTEM HEALTH AND FUNCTION

Ecosystem or ecological function is a level of coherent and adaptive organization at the ecosystem level with healthy components of water, air, soil, and diverse biota. It is the underlying structure on which agriculture is built since it regulates climate and supports essential biological processes. The agro-ecosystem, the intersection between the natural ecosystem and farming, is at the root of agricultural sustainability and must be well functioning at the ecosystem level of organization to provide services such as clean water, biodiversity, pollination, and storm protection (Costanza et al. 1997; de Groot, Wilson, and Boumans 2002; Fischer, Byerlee, and Edmeades 2009; MA 2005a).

Biodiversity is identified by many authors as a significantly degraded ecosystem service (MA 2005a; Rockstrom et al. 2009). Biologists lament the loss of biodiversity: the present rate of extinction is 1,000 times more than any such event in the fossil record (MA 2005b). Similarly, as described briefly in the first section, agriculture has experienced a decrease in diversity as crop rotations have become simpler (see Table 9.1) and livestock operations have become larger and more uniform. For crops, this trend is illustrated in Table 9.1 using the Shannon index, commonly used by ecologists to evaluate species richness and diversity. More diverse cropping landscapes have a Shannon index of close to 4, while the more uniform landscapes have values closer to 1 (Table 9.1). Also shown in Table 9.1 is the Shannon evenness index, which describes the equality of species in a population; a value of 1 describes a population with all species present in equal quantities, while a value closer to 0 represents a few species that are predominant in a population of many (Magurran 2004).

Both Canada-wide and in the prairie provinces, crop diversity has increased since 1908 overall as well as over the past three decades (Table 9.1). However, in Ontario since 1982, crop diversity has decreased and is now similar to that of 1908. In Ontario, corn-soybean-wheat rotations are common, and an upward trend in shorter rotations (corn-soybean or continuous corn or soybean) has been evident since the mid-1980s (Eilers et al. 2010). A closer look at Statistics Canada's principal field crop data for Ontario shows that, while the area seeded to corn has remained fairly static since 1982 (7 percent increase), soybean production has increased by 684,000 hectares (188 percent), mostly at the expense of mixed grains, barley and oats, and, to a lesser extent, hay (calculated from CANSIM 2013). Trading areas seeded to cereals (frequently interseeded with forages) for a more input-intensive row crop (soybean) indicates that rotations in Ontario are simplifying to intensify cash crop production. Although crop rotation is an accepted method of pest control and a promoter of soil health, farmers are often driven to intensify and simplify rotations because of financial necessity (Knowler and Bradshaw 2007)—the short-term cash benefit of a cash

Figure 9.3. A diverse agricultural landscape

SEEDED AREA OF MAJOR FIELD CROPS AS PERCENTAGE OF TOTAL

		1908	1982	1992	2002	2012	PERCENT CHANGE
PRAIRIES	DIVERSITY	1.119	1.687	1.678	1.922	1.781	+5.56
	EVENNESS	0.0738	0.0973	0.0738	0.1111	0.1034	+6.26
ONTARIO	DIVERSITY	1.703	1.912	1.881	1.815	1.682	-12.00
	EVENNESS	0.1121	0.1272	0.1254	0.1205	0.1117	-12.16
CANADA	DIVERSITY	1.760	1.951	1.947	2.236	2.125	+8.92
	EVENNESS	0.1087	0.1115	0.1111	0.1277	0.1218	+9.23

Table 9.1. Comparison of Shannon diversity index and Shannon evenness index of field crops in the Prairies, Ontario, and Canada. Source: Calculations made from seeded area (ha) in CANSIM data table Table 001-0010, Statistics Canada.

CLASS	DESCRIPTION
	——————— DEPENDABLE ———————
1	No significant limitations in use for crops
2	Moderate limitations that restrict the range of crops or require moderate conservation practices
3	Moderately severe limitations that restrict the range of crops or require special conservation practices
	——————— NON-DEPENDABLE ———————
4	Severe limitations that restrict the range of crops or require special conservation practices
5	Very severe limitations that restrict the ability to produce perennial forage crops, and improvement practices are not feasible
6	Capable only of producing perennial forage crops, and improvement practices are not feasible
7	No capacity for arable culture or permanent pasture
	——————— ORGANIC ———————
0	Organic soils (not placed in capability classes)

Table 9.2. Canadian Land Inventory land capability classes for agriculture

crop each year alleviates farm debts and low incomes. Simplification of crop rotation is one of the major areas in which mainstream agriculture has ignored agro-ecosystem health: monoculture and short rotations have decreased the proportion of perennial, cereal, and legume crops (as cover crops or green manures). These latter crops can improve soil structure and increase SOM, while legumes in particular can add soil nitrogen. Removing small grains, forages, and legumes from crop rotations decreases the soil's ability to act as a buffer. Poor soil ecosystem resilience can be more pronounced if the bulk of crop residues (which, in simple systems, are the main contributors of SOM) are removed for bio-energy feed stock or other uses.

To preserve a functional wild ecosystem on a national scale, Canada boasts nearly a million square kilometres of protected land (FPTGC 2010), but some argue that this is not sufficient to sustain biodiversity and ecosystem services. The practice of setting aside large tracts of land for wild function while intensifying farming practices on less land has been promoted as "land sparing" by ecologists, a way to produce more food while maintaining climate-regulating ecosystem function on a large scale (Green et al. 2005). However, more recent assessments conclude that, to a certain extent, "land sharing," in which farming leaves room for wild ecological function on farms, is more realistic and must form part of the solution (Tscharntke et al. 2012). Probably, both are required, in part because relying on the practice of land sparing has actually resulted in relatively small land reserves for wild ecological function and large areas of intensively farmed land.

On farms, agro-ecosystem health can be preserved by taking care of the soil (increasing soil organic matter, keeping the soil covered, using perennials, especially on marginal land); providing set-aside areas such as shelterbelts, hedgerows, windbreaks, woodlots, and riparian zones; and increasing commodity diversity.

The scope of challenges listed in Deen et al. (2013), the uncertain nature of climate variation over the next five to ten decades, and mounting evidence in the academic literature (Rockstrom et al. 2009) indicate that it is time to face the lack of attention to ecological function in agriculture by improving agro-ecosystem health. The agro-ecosystem requires more resilience to support agricultural production as we confront climate change and increasing global demand for agricultural products.

SYNERGISTIC STRATEGIES

Many solutions that can be implemented to increase resilience in the agricultural system address both agro-ecosystem health and the need to adapt to climate change. These solutions can create transformative change in the agricultural system and are described in the following sections.

Dependable and Non-Dependable Land Use – Canada's land mass is only 5 percent dependable agricultural land (Class 1, 2, and 3 land; CLI 2008; see Table 9.2)—a limited resource. With increasing pressure to maintain or increase food production under variable conditions, it is important to maintain adequate farmland that is dependable for food, fibre, fuel, and feed production. Viewing farmland as "dependable" or "non-dependable" requires looking at these issues in a slightly new light. Non-dependable land is usually termed "marginal" land because of its lack of potential for high-intensity production. We believe that "non-dependable" is a better term for it since it is less likely to sustain year-to-year stable yields. In addition, nutrient discharge from non-dependable land in poor years will be greater than on land with good soil structure, thus impacting the surrounding environment (Eilers et al. 2010). Nevertheless, Canada currently depends on Class 4 land: cultivated land has been in excess of available dependable agricultural land in Canada since the mid-1990s (Hofmann, Filoso, and Schofield 2005). This has sparked concern about the risk of soil degradation on non-dependable land used for crops.

Another distinction that can be made among the land-sharing and land-sparing arguments mentioned already (Green et al. 2005; Tscharntke et al. 2012) is what type of land should be used for each. Dependable agricultural land can maintain crop yields when intensively managed, albeit by also reducing tillage and including cover crops. These soils more likely will support consistently high yields and can be farmed while leaving wild areas for ecological function. Soils on other land classes (4–7) are non-dependable and should be more extensively managed in a land-sharing arrangement, promoting agro-ecosystem health and ecological function on a landscape scale. Row crops should not be placed on non-dependable land because the soil will support good yields primarily in ideal growing conditions. Since ideal conditions are expected to be less consistent with increasing climate variability, the use of non-dependable land might be less profitable over the next five to ten decades and more dangerous for the agro-ecosystem.

Environmental farm plans (EFPs) should incentivize not only stewardship efforts on farmland in general but also encourage additional efforts on non-dependable land and sensitive areas prone to erosion. The use of perennial forages on non-dependable land is ideal for managed grazing and bringing crop and livestock production closer together to minimize nutrient discharge into the environment.

Conservation of Agricultural Land – Since 1951, Canada has lost 1.8 million hectares of dependable agricultural land to urbanization, transportation and utility infrastructure, and rural housing, the worst of which has occurred in southern Ontario (Hofmann, Filoso, and Schofield 2005). The loss of agricultural land

RECOMMENDATION	SUPPORTED BY
Incorporate wheat, forages, and legumes into corn-soybean rotations	
Build SOM/increase soil health	
	Thorough, widely taken-up EFPs
Implement and encourage conservation practices and set-aside areas on farms	
Appropriate use of non-dependable land	
Develop and make available hardy, resistant crop varieties	Plant-breeding research
Conserve agricultural land	Agricultural land reserves and regulation
Integrate crop and livestock systems	Policy regulating large farms and consumer choice

Table 9.3. Recommendations and required support mechanisms

in southern British Columbia caused initiation of the Agricultural Land Reserve in 1973, protecting over 4.5 million hectares, while the same problem in Ontario resulted in the creation of the Ontario Greenbelt in 2005. The Greenbelt protects 720,000 hectares of land around the fast-growing Greater Golden Horseshoe area (Friends of the Greenbelt 2012). As conditions become more variable and year-to-year regional yield stability becomes more important, conserving dependable agricultural land is crucial. Keeping land in agricultural use can stimulate the economy by providing an economic multiplier effect of up to three times direct earnings from agriculture (Francis et al. 2012). The revenue multiplier effect of agriculture is larger than that of any other industry, meaning that agricultural activity in a given area increases the profitability of other industries more than any other kind of economic activity (Cross and Ghanem 2006). However, this is not always apparent when stakeholders are faced with decisions, such as the development of agricultural land, that can be more immediately profitable (Francis et al. 2012).

A few strategies can be applied to conserve agricultural land. Strong initiatives such as the Agricultural Land Reserve in the Lower Mainland of British Columbia and the Ontario Greenbelt in the Greater Golden Horseshoe need to be reinforced, expanded, and emulated (Friends of the Greenbelt 2012; Smith and Haid 2004). Where urban development is allowed to expand into farmland, it should do so on non-dependable land in a way that supports agricultural and food systems, for example by including facilities for market gardening and local food processing. Land trust programs also exist across the country and warrant support. Examples are the Ontario Farmland Trust, the Genesis Land Conservancy in Saskatchewan, the New Brunswick Community Land Trust, and the Land Conservancy of BC (Caldwell and Hilts 2005; Gorsuch and Scott 2010).

Maintain or Increase Soil Health – Soil is a complex ecosystem and requires careful management to continue being productive year after year. Soil organic matter is a central element of sustainable soil stewardship. The organic portion of soil is made up of sugars, proteins, lipids, and humic substances and contains carbon, nitrogen, oxygen, hydrogen, phosphorous, and sulfur. Soil organic carbon (SOC) is the carbon content of the soil and is used as a measure of SOM (Acton and Gregorich 1995; Brady and Weil 1996; Kibblewhite, Ritz, and Swift 2008). SOM moderates soil function by providing aeration, increasing water infiltration, holding soil moisture, acting as a substrate for soil biota, and contributing to fertility and stability (Brady and Weil 1996). It is a balancing act to maintain active SOM, which breaks down and releases nutrients necessary for crop growth, while maintaining passive SOM, which sequesters carbon and can contribute to soil structure (Janzen 2006; Loveland and Webb 2003). The active fraction is quickly depleted, provided that there is sufficient N, if not "fed" by adding carbon-based material: crop residues. In simple rotations, fertilizer N can maintain or increase SOC over time when it increases crop residue return (Gregorich, Drury, and Baldock 2001; Loveland and Webb 2003). The effect is greater when manure and/or a forage rotation is added (Hofmann et al. 2009; Ladha et al. 2011). For example, VandenBygaart, Gregorich, and Angers (2003) summarize the literature indicating that the net SOC gain from adding legumes in rotation to continuous corn could be 14.4 tonnes per hectare per year. Especially in systems dependent on synthetic inputs, it is not prudent to remove crop residues crucial for maintaining SOM (Loveland and Webb 2003).

Because soil that is high in SOM is better able to provide a conducive growing environment, it also acts as a buffer to biotic (pest, weed, and disease challenges) and abiotic (water excess or deficit, high or low temperature) shocks, which might become more frequent with climate change. Key strategies for increasing SOM are (1) retaining crop residues (Carefoot, Janzen, and

Lindwall 1994) and (2) diversifying crop rotation, especially by adding deep-rooted perennial forages Gaudin et al. 2013; Vanden Bygaart et al. 2010).

In Canada, soils are generally in "good health," as defined by Environment Canada (Eilers et al. 2010). Nine percent of Canada's cultivated land lost more than eleven tonnes of soil per hectare per year in 2006, much of it because of water erosion, while 80 percent lost less than six tonnes of soil per hectare per year (ibid.). Improved practices over the years in response to the Dust Bowl conditions of the 1930s have been important for soil conservation across the North American Great Plains (Baveye et al. 2011). The western provinces have taken steps to mitigate the effects of wind erosion, to which their land is susceptible, by changing practices away from summerfallow and toward conservation or reduced tillage (a 36 percent increase in land prepared with conservation tillage from 2001 to 2011; calculated from CANSIM 2013). In the prairie provinces, this is because of rotations that are more complex (less use of fallow, thus less soil degradation and more production on the same amount of land) and diversified (more pulses, oilseeds, and forages; Acton and Gregorich 1995; Padbury et al. 2002; Statistics Canada 2013). Since 1976, prairie summerfallow land decreased 80 percent to about 2 million hectares fallowed in 2011, and over 80 percent of prepared land was under conservation tillage (calculated from CANSIM 2013). Soils under conservation management are expected to better withstand intense rainfall events and have more tolerance to droughts (Govaerts et al. 2009; Kibblewhite, Ritz and Smith 2008).

Soil erosion is more prevalent in eastern Canada and is mostly associated with intensive crops such as potato, sugar beet, and horticulture, in short crop rotations. Unfortunately, conservation tillage, so effective in the west, does not hold the same promise for eastern Canada because of factors such as higher levels of precipitation and differences in tillage and cropping (Angers et al. 1997; VandenBygaart, Gregorich, and Angers 2003). Some soils, particularly in the Atlantic provinces and southern Ontario, have low and decreasing soil organic carbon (Eilers et al. 2010). From 1981 to 2006, the percentage of Ontario soils that lost more than twenty-two tonnes of soil per hectare per year remained fairly static at 46 percent in 1981 and 41 percent in 2006 (ibid.). This situation requires attentive management especially under intensive row crops susceptible to water erosion (Miller 1985). Row crop systems often result in bare soil between rows and during non-growing seasons. Solutions include adding cover crops to the rotation, which reduce erosion during intense precipitation events outside the growing season. Intercropping, terracing on slopes, and adding hedgerows are strategies that can be useful during the growing season. Non-dependable soils (Classes 4–7; CLI 2008) should not be used for intensive row crops to avoid soil erosion and degradation.

Moving toward long-term sustainability of Canadian soils, the east needs to implement practices to control soil degradation using emerging strategies and longer rotations, particularly including forages. When forages and manuring are incorporated into cropping systems, SOM increases, indicating that livestock-crop integration can be a beneficial soil stewardship practice (Davis et al. 2012). Concentrating on residue cover and return, cover crops and intercropping, addition of manure, and perennial forages to protect soil from erosion will ensure that agricultural soils can continue to provide food, feed, fuel, fibre, and ecosystem services in the future. Effective soil management is less dependent on the farming system (conventional, organic) than on the management of crop residues, soil amendments, cover crops, and forages in rotation (Nelson 2005).

Increase Complexity – As described, agricultural and natural landscapes have become simpler over the past 100 years. Diversity is an important element of resilience (FSPG 2008; MA 2005b) and incorporates genetic diversity, crop diversity, on-farm diversity, wild species diversity, and commodity diversity. The experience of the Canadian prairies is an excellent example: by decreasing summerfallow and increasing the complexity of crop rotations, there has been an increase in soil water-holding capacity, while air pollution by dust has decreased (Acton and Gregorich 1995; Eilers et al. 2010; Padbury et al. 2002). Not only has this increased the health of the agro-ecosystem by influencing soil, water, and air quality, but also it has decreased vulnerability to drought.

Ontario and eastern Canada have not witnessed an increase in rotation complexity—in fact, as described, crop rotation diversity has decreased (Table 9.1). The most common rotation includes corn, soybean, and sometimes wheat. OMAFRA chronicles the difficulty of encouraging producers to continue using wheat in rotations because of the higher commodity prices of corn and soybean (Johnson 2010). However, Deen et al. (unpublished data) found that more complex rotations in Ontario led to increased stability of year-to-year yields. In drought years, the rotations with wheat performed better than corn or corn-soy rotations. This effect is invaluable when we are looking to adapt our systems to increased weather variability.

To take the next step in year-to-year yield stability, producers should add a forage crop to their rotations, particularly a forage legume or legume mix. These crops can be grazed, harvested for hay or silage, or plowed under to provide nitrogen and improved soil structure for following crops. Legumes' ability to fix nitrogen decreases the need for nitrogen fertilizer application, and as a perennial cover crop they limit soil erosion—a problem in eastern Canada. The benefits of adding forage legumes in rotations include increasing water-holding capacity and decreasing nutrient loss to the environment, as discussed above in the section on soil health. In a recent paper, Gaudin et al. (2013) suggest that red clover is the ultimate addition to the corn-soy-wheat rotation in Ontario. Alfalfa is another forage legume well suited to rotations with corn (Gregorich, Drury, and Baldock 2001).

In a British study, Abson, Fraser, and Benton (2013) indicate that an increased commodity diversity at the landscape scale (each region studied had an area of 578 square kilometres) increases stability of profit margins across years. Increasing diversity on a landscape scale also closes nutrient cycles. If livestock and crops are produced in the same area, there is more land available on which to spread manure at reasonable transportation costs (Jackson, Keeney, and Gilbert 2000; Zebarth et al. 1998). This should decrease nutrient discharge into waterways and groundwater from the application of more nutrients than crops can take up. Russelle, Entz, and Fransluebbers (2007) explain in detail which options exist for integrating livestock and crop production, including both on the farm and regionally. A study from Simon Fraser University (Karlsen and Newman 2013) indicates that large beef feedlots can become unsustainable with increasing climate variability because of variability in water and grain availability, leading to a transition back to smaller operations in which the cattle are grass fed. This would be a decentralization of beef production, allowing for closer nutrient cycling, increasing commodity diversity with a positive effect on the local economy, and minimizing the risk to beef operations in case of floods and droughts.

In Britain, this decentralization has already started to some extent. Farms have decreased in size partly because of the Single Farm Payment program, which issues payments depending on management for soil conservation, SOM, and wild habitat conservation (Angus et al. 2009). In short, farmers are being paid to adapt to climate change, and the payment system has encouraged more small farms and diversified the farm landscape.

We suggest that, to address the problem of nutrient discharge from large farms and improve adaptive ability to climate change simultaneously, livestock operations would become smaller and more discrete—more integrated with cropping systems so that nutrient cycling (from soil to feed to animal and back to soil) is made more efficient and on a smaller scale. This model would also allow for greater use of forages and small grain cereals in ruminant livestock production (Oltjen and Beckett 1996), thus freeing grains such as corn and soybean for direct human consumption. Forages also increase soil health (Davis et al. 2012). Modifying the agricultural production system so that livestock and crops are more closely integrated and more forages are used is a transformative change that would boost the health of the agro-ecosystem and increase the agriculture system's resilience in the face of climate change.

Moving away from highly concentrated livestock production will decrease the availability of cheap animal protein. Pimentel and Pimentel (2003) explain the debate between food and feed succinctly and conclude that animal protein is less efficient in terms of energy, land, and water. Transformative change to decrease the use of human-edible food products as animal feed could substantially increase the quantity of food available. Consumption of edible food (mostly corn and soybean) by livestock allows us to exploit economies of scale, but ruminants in particular are more suited to eat grass, forage legumes, and small grains (Oltjen and Beckett 1996). Forage-based livestock systems certainly require more land ("non-dependable" or otherwise). A change in favour of more integrated, forage-based livestock systems would require a decrease in the amount of meat consumed by people in developed countries (Aiking 2011; Pimentel and Pimentel 2003).

One impact of decreased biodiversity is a loss of genetic information in crops and livestock, which threatens our ability to develop novel, adapted varieties. There is a significant movement to maintain diverse seed stocks in seed banks worldwide (Charles 2006), and there are many initiatives in place to preserve genetically important livestock breeds (Notter 1999). In the interest of supporting diversity for resilience, it is important to recognize that alternative agricultural systems, often dismissed by mainstream agriculture ("organic can't feed the world"), have a substantial part to play in maintaining a strong food system over the next few hundred years. Alternative farming acts as a reservoir for traditional breeds and varieties and a testing ground for adaptive ideas. Diversity of farming systems is yet another way to increase resilience by widening the base on which agriculture depends.

Maintaining Conserved Areas on the Farm – To make farms more resilient, to decrease wind damage, and to increase water and nutrient holding capacity, more set-aside areas are required on farms. With rising land prices, farmers might be driven to remove grassed waterways, fence rows, hedgerows, and windbreaks, and decrease the size of riparian zones. But these features might help farms to adapt to climate change and should be incentivized. With flooding, the flushing of nutrients and sediment into waterways decreases the chances of the farm bouncing back quickly. Leaving cropped land more exposed by removing hedgerows and windbreaks could contribute to more wind erosion and decrease soil fertility and organic matter, which, as we have already seen, decrease adaptive ability.

Canadian farms used some conservation practices in 2006: 34 percent of farmers used perennial forages on marginal land, 23 percent used cover and companion crops, 60 percent had riparian setbacks, but only 32 percent had limited livestock access to waterways (Eilers et al. 2010). Currently, the EFP is quite effective in promoting the use of some conservation strategies, such as riparian zones (AAFC 2012; Rajsic, Ramlal, and Fox 2012) but does not suffi-ciently encourage other conserved areas on the farm. In the United States, the Conservation Reserve Program has encouraged producers to take 12 million hectares of land out of high-intensity production to contribute to a resilient ecosystem and to provide a reservoir of land for future production, should it be required (Francis et al. 2012). These strategies need to be enhanced to diver-sify the ubiquitous corn-soybean crops that dominate so much of the eastern Canadian agricultural landscape. A transformation is required; conservation practices need to be used on 100 percent of farms.

Develop Hardy Crop Varieties – Breeding efforts focused on hardy crops with effi-cient nitrogen utilization and resistance to droughts, floods, new pest challenges, and pollutants such as ozone are important (e.g., Comeau et al. 2010). For this to occur, it is necessary to invest in educating and funding the work of plant breed-ers (Tester and Landridge 2010; Varshney et al. 2011). A transition to more hardy varieties might involve movement away from the approach of maximizing crop yields. Instead, a focus on optimizing or stabilizing yields across years and re-gions could be taken, perhaps using a rolling average to describe yields over time in a given area. An approach that takes into account the increased variation in climatic events, using varieties adapted to higher atmospheric levels of CO_2 and ozone, droughts, and floods and less dependent on energy-intensive inputs (e.g., nitrogen fertilizer), will increase resilience in farming over the next 100 years.

LOOKING AHEAD WITH AGRO-ECOSYSTEM HEALTH AND CLIMATE CHANGE ADAPTATION
Going forward, policy can usefully support research for integrated agricultural systems, strategies for improving soil health and conservation on farms, and new crop varieties. Diversity at all levels is imperative to foster resilience to shocks to the agricultural system.

As farms and the farm landscape adapt to climate change with more diverse and complex systems, increased heterogeneity of operations, closed nutrient cycles, improved soil stewardship, and hardier varieties, rural com-munities can be revitalized. Depending on a broader base for agricultural production means that more rural areas will be involved. More resilience in agriculture could mean, in part, more resilience in rural living, encouraging vibrant, diverse communities. Some suggest that a viable, sustainable farming system has as much to do with the quality of life of farmers and their neigh-bours as it does with production and agro-ecosystem health (Russelle, Entz, and Fransluebbers 2007; Scott 2006).

Notes

1   Cultivated land; defined by Statistics Canada as total land in crops, summerfallow, and tame hay (CANSIM 2013).

2   Land in crops; defined by Statistics Canada as the sum of land used for cultivation of field crops, vegetables, fruits, nursery crops, and sod (CANSIM 2013).

References

Abson, D.J., E.D.G. Fraser, and T.G. Benton. 2013. "Landscape Diversity and the Resilience of Agricultural Returns: A Portfolio Analysis of Land Use Patterns and Economic Returns from Lowland Agriculture." *Agriculture and Food Security* 2: 2–15.

Acton, D.F., and L.J. Gregorich, eds. 1995. *The Health of Our Soils: Toward Sustainable Agriculture in Canada.* Centre for Land and Biological Resources Research, Research Branch, Agriculture and Agri-Food Canada, Catalogue No. A53-1906/1995E. Ottawa.

Adamo, S.A., J.L. Baker, M.M.E. Lovett, and G. Wilson. 2012. "Climate Change and Temperate Zone Insects: The Tyranny of Thermodynamics Meets the World of Limited Resources." *Evironmental Entomology* 41: 1644–52.

Agriculture and Agri-food Canada. 2012. *An Overview of the Canadian Agriculture and Agri-Food System 2012.* Research and Analysis Directorate, Strategic Policy Branch, Catalogue No. A38-1/1-2011E-PDF.

Aiking, H. 2011. "Future Protein Supply." *Trends in Food Science Technology* 22: 112–20.

Angers, D.A., M.A. Bolinder, M.R. Carter, E.G. Gregorich, C.F. Drury, B.C. Liang, R.P. Voroney, R.R. Simard, R.G. Donald, R.P. Beyaert, and J. Martel. 1997. "Impact of Tillage Practices on Organic Carbon and Nitrogen Storage in Cool, Humid Soils of Eastern Canada." *Soil Tillage Research* 41: 191–201.

Angus, A., P.J. Burgess, J. Morris, and J. Lingard. 2009. "Agriculture and Land Use: Demand for and Supply of Agricultural Commodities, Characteristics of the Farming and Food Industries, and Implications for Land Use in the UK." *Land Use Policy* 26S: S230–42.

Baveye, P-C., D. Rangel, A.R. Jacobson, M. Laba, C. Darnault, W. Otten, R. Radulovich, and F.A.O. Camargo. 2011. "From Dust Bowl to Dust Bowl: Soils Are Still Very Much a Frontier of Science." *Soil Science Society of America Journal* 75: 2037–48.

Brady, N.C., and R.R. Weil. 1996. *The Nature and Properties of Soils.* 11th ed. New Jersey: Prentice-Hall.

Caldwell, W., and S. Hilts. 2005. "Farmland Preservation: Innovative Approaches in Ontario." *Journal of Soil and Water Conservation* 60: 66A–69A.

CANSIM. 2013. Statistics Canada. http://www5.statcan.gc.ca/cansim/a01?lang=eng.

Carefoot, J.M., H.H. Janzen, and C.W. Lindwall. 1994. "Crop Residue Management for Irrigated Cereals on the Semi-Arid Canadian Prairies." *Soil Tillage Research* 32: 1–20.

Charles, D. 2006. "A 'Forever' Seed Bank Takes Root in the Arctic." *Science* 312: 1730–31.

Chesnaux, R., D.M. Allen, and G. Graham. 2007. "Assessment of the Impact of Nutrient Management Practices on Nitrate Contamination in the Abbotsford-Sumas Aquifer." *Environmental Science and Technology* 41: 7229–34.

CLI (Canada Land Inventory), 2008. "Overview of Classification Methodology for Determining Land Capability for Agriculture." Agriculture and Agri-Food Canada. http://sis.agr.gc.ca/cansis/nsdb/cli/classdesc.html.

Comeau, A., L. Nodichao, J. Collin, M. Baum, J. Samsatly, D. Hamidou, F. Langevin, A. Larouche, and E. Picard. 2010. "New Approaches for the Study of Osmotic Stress Induced by Polyethylene Glycol (PEG) in Cereal Species." *Cereal Research Communications* 38: 471–81.

Costanza, R., R. d'Arge, R. de Groot, S. Farber, M. Grasso, B. Hannon, K. Limburg, S. Naeem, R.V. O'Neill, J. Paruelo, R.G. Raskin, P. Sutton, and M. van den Belt. 1997. "The Value of the World's Ecosystem Services and Natural Capital." *Nature* 387: 253–60.

Cross, P., and Z. Ghanem. 2006. "Multipliers and Outsourcing: How Industries Interact with Each Other and Affect GDP." *Canadian Economic Observer* 19, 1: 18–35. Statistics Canada Catalogue No. 11-010-XIB.

Davis, A.S., J.D. Hill, C.A. Chase, A.M. Johanns, and M. Liebman. 2012. "Increasing Cropping System Diversity Balances Productivity, Profitability, and Environmental Health." *PLOS ONE* 7: 1–8, e47149.

de Groot, R.S., M.A. Wilson, and R.M.J. Boumans. 2002. "A Typology for the Classification, Description, and Calculation of Ecosystem Functions, Goods, and Services." *Ecological Economics* 41: 393–408.

Deen, B., M.E. Graves, E.D.G. Fraser, and R.C. Martin. 2013. "Changing Demands on Agricultural Land: Are Reforms Urgent?" Green paper for the Alberta Institute of Agrologists. http://www.albertaagrologists.ca/default.aspx?page=802.

Drury, C.F., and C.S. Tan. 1995. "Long-Term (35 Years) Effects of Fertilization, Rotation, and Weather on Corn Yields." *Canadian Journal of Plant Science* 75: 355–62.

Dukes, J.S., J. Pontius, D. Orwig, J.R. Garnas, V.L. Rodgers, N. Brazee, B. Coke, K.A. Theoharides, E.E. Stange, R. Harrington, J. Ehrenfeld, J. Gurevitch, M. Lerdau, K. Stinson, R. Wick, and M. Ayres. 2009. "Responses of Insect Pests, Pathogens, and Invasive Plant Species to Climate Change in the Forests of Northeastern North America: What Can We Predict?" *Canadian Journal of Forestry Research* 39: 231–48.

Eilers, W., R. MacKay, L. Graham, and A. Lefebvre, eds. 2010. *Environmental Sustainability of Canadian Agriculture: Agri-Environmental Indicator Report Series—Report #3.* Ottawa: Agriculture and Agri-Food Canada.

Evans, D.O., K.H. Nichols, Y.C. Allen, and M.J. McMurtry. 1996. "Historical Land Use, Phosphorus Loading, and Loss of Fish Habitat in Lake Simcoe, Canada." *Canadian Journal of Fisheries and Aquatic Sciences* 53: 194–218.

Fischer, R.A., D. Byerlee, and G.O. Edmeades. 2009. "Can Technology Deliver on the Yield Challenge to 2050?" Presented to the FAO Expert Meeting on How to Feed the World in 2050, Rome.

FPTGC (Federal, Provincial, and Territorial Governments of Canada). 2010. *Canadian Biodiversity: Ecosystem Status and Trends 2010.* Ottawa: Canadian Councils of Resource Ministers.

Francis, C.A., T.E. Hansen, A.A. Fox, P.J. Hesje, H.E. Nelson, A.E. Lawseth, and A. English. 2012. "Farmland Conversion to Non-Agricultural Uses in the US and Canada: Current Impacts and Concerns for the Future." *International Journal of Agricultural Sustainability* 10: 8–24.

Friends of the Greenbelt. 2012. "Facts and Figures." http://greenbelt.ca/about-greenbelt/facts-figures.

FSPG (Food Security Policy Group). 2008. "Pathways to Resilience: Smallholder Farmers and the Future of Agriculture." http://foodgrainsbank.ca/resilience.aspx.

Gal, A., T.J. Vyn, E. Micheli, E.J. Kladivko, and W.W. McFee. 2007. "Soil Carbon and Nitrogen Accumulation with Long-Term No-Till versus Moldboard Plowing Overestimated with Tilled-Zone Sampling Depths." *Soil Tillage Research* 96: 42–51.

Gaudin, A.C.M., S. Westra, C.E.S. Loucks, K. Janovicek, R.C. Martin, and W. Deen. 2013. "Improving Resilience of Northern Field Crop Systems Using Inter-Seeded Red Clover: A Review." *Agronomy* 3: 148–80. doi:10.3390/agronomy3010148.

Gornall, J., R. Betts, E. Burke, R. Clark, J. Camp, K. Willett, and A. Wiltshire. 2010. "Implications of Climate Change for Agricultural Productivity in the Early Twenty-

First Century." *Philosophical Transactions of the Royal Society, Biological Sciences* 365: 2973–89.

Gorsuch, W., and R. Scott. 2010. *A Review of Farmland Trusts: Communities Supporting Farmland, Farming, and Farmers.* Victoria, BC: Land Conservancy of BC. http://www.farmfolkcityfolk.ca/PDFs_&_Docs/CFPdocs/FLT_web.pdf.

Govaerts, B., N. Verhulst, A. Castellanos-Navarrete, K.D. Sayre, J. Dixon, and L.Dendooven. 2009. "Conservation Agriculture and Soil Carbon Sequestration: Between Myth and Farmer Reality." *Critical Reviews in Plant Science* 28: 97–122.

Great Lakes Commission. 2012. *Priorities for Reducing Phosphorus Loading and Abating Algal Blooms in the Great Lakes–St. Lawrence River Basin: Opportunities and Challenges for Improving Great Lakes Aquatic Ecosystems.* Ann Arbor: Phosphorus Reduction Task Force. http://www.glc.org/announce/12/pdf/FINAL_PTaskForceReport_Sept2012.pdf.

Green, R.E., S.J. Cornall, J.P.W. Scharlemann, and A. Balmford. 2005. "Farming and the Fate of Wild Nature." *Science* 307: 550–55.

Gregorich, E.G., C.F. Drury, and J.A. Baldock. 2001. "Changes in Soil Carbon under Long-Term Maize in Monoculture and Legume-Based Rotation." *Canadian Journal of Soil Science* 81: 21–31.

Hayes, F., G. Mills, L. Jones, and M. Ashmore. 2010. "Does a Simulated Upland Grassland Community Respond to Increasing Background, Peak, or Accumulated Exposure of Ozone?" *Atmospheric Environment* 44: 4155–64.

Hofmann, A., A. Heim, P. Giocchini, A. Miltner, M. Gehre, and M.W.I. Schmidt. 2009. "Mineral Fertilization Did Not Affect Decay of Old Lignin and SOC in a 13C-Labeled Arable Soil over 36 Years." *Biogeosciences* 6: 1139–48.

Hofmann, N., G. Filoso, and M. Schofield. 2005. "The Loss of Dependable Agricultural Land in Canada." *Rural and Small Town Canada Analysis Bulletin* 6: 1–16. Statistics Canada Catalogue No. 21-006-XIE.

IPCC (Intergovernmental Panel on Climate Change). 2007. "Summary for Policymakers." In *Climate Change 2007: Impacts, Adaptation, and Vulnerability,* edited by M.L. Parry, O.F. Canziani, J.P. Palutikof, P.J. van der Linden, and C.E. Hanson, 7–22. Contribution of Working Group II to the Fourth Assessment Report of the Intergovernmental Panel on Climate Change. Cambridge, UK: Cambridge University Press.

Jackson, L.L., D.R. Keeney, and E.M. Gilbert. 2000. "Swine Manure Management Plans in North-Central Iowa: Nutrient Loading and Policy Implications." *Journal of Soil and Water Conservation* 55: 205–12.

Jaggard, K.W., A. Qi, and E.S. Ober. 2010. "Possible Changes to Arable Crop Yields by 2050." *Philosophical Transactions of the Royal Society, Biological Sciences* 365: 2835–51.

Janzen, H.H. 2006. "The Soil Carbon Dilemma: Shall We Hoard It or Use It?" *Soil Biology and Biochemistry* 38: 419–24.

Janzen, H.H., K.A. Beauchemin, Y. Bruinsma, C.A. Campbell, R.L. Desjardins, B.H. Ellert, and E.G. Smith. 2003. "The Fate of Nitrogen in Agroecosystems: An Illustration Using Canadian Estimates." *Nutrient Cycling in Agroecosystems* 67: 85–102.

Johnson, P. 2010. "Cereals a Tough Sell … Again!" OMAFRA Croptalk Update, April. http://www.agrinewsinteractive.com/archives/article-10767.htm.

Karlsen, E., and L. Newman. 2013. *Background Report: Climate Change Adaptation and Canada's Crops and Food Supply.* Burnaby, BC: Adaptation to Climate Change Team, Simon Fraser University. http://act-adapt.org/food-supply/.

Kibblewhite, M.G., K. Ritz, and M.J. Swift. 2008. "Soil Health in Agricultural Systems." *Philosophical Transactions of the Royal Society, Biological Sciences* 363: 685–701.

Knowler, D., and B. Bradshaw. 2007. "Farmers' Adoption of Conservation Agriculture: A Review and Synthesis of Recent Research." *Food Policy* 32: 25–48.

Ladha, J.K., C.K. Reddy, A.T. Padre, and C. van Kessel. 2011. "Role of Nitrogen Fertilization in Sustaining Organic Matter in Cultivated Soils." *Journal of Environmental Quality* 40: 1756–66.

Li, Y., W. Ye, M. Wang, and X. Yan. 2009. "Climate Change and Drought: A Risk Assessment of Crop-Yield Impacts." *Climate Research* 39: 31–46.

Liebman, M.Z., M.J. Helmers, L.A. Schulte-Moore, and C.A. Chase. 2013. "Using Biodiversity to Link Agricultural Productivity with Environmental Quality: Results from Three Field Experiments in Iowa." *Renewable Agriculture and Food Systems* 28: 115–28.

Loveland, P., and J. Webb. 2003. "Is There a Critical Level of Organic Matter in the Agricultural Soils of Temperate Regions? A Review." *Soil Tillage Research* 70: 1–18.

MA (Millennium Ecosystem Assessment). 2005a. *Ecosystems and Human Well-Being: Synthesis.* Washington, DC: Island Press.

———. 2005b. *Ecosystems and Human Well-Being: Biodiversity Synthesis.* Washington, DC: World Resources Institute.

Magurran, A.E. 2004. "Chapter Four: An Index of Diversity...." In *Measuring Biological Diversity,* 100–30. Malden: Blackwell Science.

Miller, I.H. 1985. "Soil Degradation in Eastern Canada: Its Extent and Impact." *Canadian Journal of Agricultural Economics* 33: 7–18.

Nelson, A.G. 2005. "Soil Erosion Risk and Mitigation through Crop Rotation on Organic and Conventional Cropping Systems." MSc thesis, University of Manitoba.

Notter, D.R. 1999. "The Importance of Genetic Diversity in Livestock Populations of the Future." *Journal of Animal Science* 77: 61–69.

Oltjen, J.W., and J.L. Beckett. 1996. "Role of Ruminant Livestock in Sustainable Agricultural Systems." *Journal of Animal Science* 74: 1406–09.

Padbury, G., S. Waltman, J. Caprio, G. Coen, S. McGinn, D. Mortensen, G. Nielsen, and R. Sinclair. 2002. "Agroecosystems and Land Resources of the Northern Great Plains." *Agronomy Journal* 94: 251–61.

Pimentel, D., and M. Pimentel. 2003. "Sustainability of Meat-Based and Plant-Based Diets and the Environment." *American Journal of Clinical Nutrition* 78: 660S–63S.

Prather, M., D. Ehhalt, F. Dentener, R. Derwent, E. Dlugokencky, E. Holland, I. Isaksen, J. Katima, V. Kirchhoff, and P. Matson. 2001. "Atmospheric Chemistry and Greenhouse Gases." In *Climate Change 2001: The Scientific Basis,* edited by J.T. Houghton, Y. Ding, D.J. Griggs, M. Noguer, P.J. van der Linder, X. Dai, K. Maskell, and C.A. Johnson, 239–87. Contribution of Working Group I to the Third Assessment Report of the Intergovernmental Panel on Climate Change. Cambridge, UK: Cambridge University Press.

Rajsic, P., E. Ramlal, and G. Fox. 2012. "Canadian Agricultural Environmental Policy: From the Right to Farm to Farming Right." In *The Economics of Regulation in Agriculture: Compliance with Public and Private Standards,* edited by F.M. Brouwer, G. Fox, and R.A. Jongeneel, 55–78. Oxfordshire, UK: CAB International.

Rockstrom, J., W. Steffen, K. Noone, Å. Persson, F.S. Chapin III, E. Lambin, T.M. Lenton, M. Scheffer, C. Folke, H. Schellnhuber, B. Nykvist, C.A. De Wit, T. Hughes, S. van der Leeuw, H. Rodhe, S. Sörlin, P.K. Snyder, R. Costanza, U. Svedin, M. Falkenmark, L. Karlberg, R.W. Corell, V.J. Fabry, J. Hansen, B. Walker, D. Liverman, K. Richardson, P. Crutzen, and J. Foley. 2009. "Planetary Boundaries: Exploring the Safe Operating Space for Humanity." *Ecology and Society* 14: 1–32. http://www.ecologyandsociety.org/vol14/iss2/art32/.

Royal Society. 2008. *Ground-Level Ozone in the 21st Century: Future Trends, Impacts, and Policy Implications*. London, UK: Science Policy Report.

Russelle, M.P., M.H. Entz, and A.J. Fransluebbers. 2007. "Reconsidering Integrated Crop-Livestock Systems in North America." *Agronomy Journal* 99: 325–34.

Scott, J. 2006. *Can There Be a Magic Pudding? Towards an Understanding of Viable Farms*. New South Wales, Australia: Marketing Services and Publications, University of New England. National Library of Australia Catalogue No. 338.140994. http://www.une.edu.au/ers/documents/magic-pudding.pdf.

Smil, V. 2000. "Phosphorus in the Environment: Natural Flows and Human Interferences." *Annual Review of Energy and The Environment* 25: 53–88.

——. 2012. "Nitrogen Cycle and World Food Production." *World Agriculture* 2: 9–13.

Smith, B.E., and S. Haid. 2004. "The Rural-Urban Connection: Growing Together in Greater Vancouver." *Plan Canada* s: 36–39.

Smith, P., D. Martino, Z. Cai, D. Gwary, H. Janzen, P. Kumar, B. McCarl, S. Ogle, F. O'Mara, C. Rice, B. Scholes, and O. Sirotenko. 2007. "Agriculture." In *Climate Change 2007: Mitigation*, edited by B. Metz, O.R. Davidson, P.R. Bosch, R. Dave, and L.A. Meyer, 497–540. Contribution of Working Group III to the Fourth Assessment Report of the Intergovernmental Panel on Climate Change. Cambridge, UK: Cambridge University Press.

Stocker, T. 2010. *Video Message. Working Group I: The Physical Science Basis*. IPCC. http://www.ipcc.ch/news_and_events/docs/COP16/IPCC/stocker10cancun-videomessage.pdf.

Tester, M., and M. Langridge. 2010. "Breeding Technologies to Increase Crop Production in a Changing World." *Science* 327: 818–22.

Tscharntke, T., Y. Clough, T.C. Wanger, L. Jackson, I. Motzke, I. Perfecto, J. Vandermeer, and A. Whitbread. 2012. "Global Food Security, Biodiversity Conservation, and the Future of Agricultural Intensification." *Biological Conservation* 151: 53–59.

Van Horn, H.H. 2002. "Manure/Effluent Management: Nutrient Recycling." In *Encyclopedia of Dairy Sciences*, edited by H. Roginski, J.W. Fuquay, and P.F. Fox, 1714–22. London: Academic Press.

VandenBygaart, A.J., E.G. Gregorich, and D.A. Angers. 2003. "Influence of Agricultural Management on Soil Organic Carbon: A Compendium and Assessment of Canadian Studies." *Canadian Journal of Soil Science* 83: 363–80.

VandenBygaart, A.J., E. Bremer, B.G. McConkey, H.H. Janzen, D.A. Angers, M.R. Carter, C.F. Drury, G.P. Lafond, R.H. McKenzie. 2010. "Soil Organic Carbon Stocks on Longterm Agroecosystem Experiments in Canada." *Canadian Journal of Soil Science* 90: 543–550.

Varshney, R.K., K.C. Bansal, P.K. Aggarwal, S.K. Datta, and P.Q. Craufurd. 2011. "Agricultural Biotechnology for Crop Improvement in a Variable Climate: Hope or Hype?" *Trends in Plant Science* 16: 363–71.

Zebarth, B.J., B. Hii, H. Liebscher, K. Chipperfield, J.W. Paul, G. Grove, andS.Y. Szeto. 1998. "Agricultural Land Use Practices and Nitrate Contamination in the Abbotsford Aquifer, British Columbia, Canada." *Agriculture, Ecosystems and Environment* 69: 99–112.

CHAPTER TEN

RURAL SUSTAINABILITY

CHRISTOPHER BRYANT

In this final chapter, I begin with a discussion of rural sustainability and what it means for the sustainable development of rural systems. Next I discuss the main forces affecting rural systems and communities that have to be taken into account in planning the way forward, including the many conflicts and uncertainties that they present. This discussion contributes to the following section, which explains how these forces underlie much of the vulnerability of different rural systems, their territories, and their communities. This section is followed by a discussion of community resilience in relation to the multitude of forces affecting rural systems. A thread running throughout is the importance of reducing vulnerability and building community resilience *before* difficult circumstances or disasters in one form or another have to be confronted.

THE SUSTAINABLE DEVELOPMENT OF RURAL SYSTEMS, THEIR COMMUNITIES, AND THEIR ACTIVITIES

Rural sustainability is explored in terms of its implications for the development of rural territories and communities and their activities and populations. First, sustainable development refers to its three commonly identified essential dimensions: namely, the environmental, social, and economic dimensions. Sustainable development is thus the search for development that is environmentally friendly, socially acceptable, and economically viable. Second, in addition to these three dimensions, a fourth dimension has become increasingly recognized as being so important in the search for sustainable solutions and sustainable development generally that it must be identified explicitly, even though it can also be seen as part of the social dimension. This is the dimension of governance, implying the need to focus on processes of management and planning, including assuming responsibility for taking and managing initiatives.

Furthermore, sustainable development is difficult to identify as a specific end result of development processes because all three of the basic dimensions can change over time. The *environmental* dimension changes as our understanding of different aspects of the natural environment improves and therefore changes. An important example for this book is what is happening to climate and its variability. Although we can observe what is happening today and in the recent past, our understanding of the dynamics of climate into the future is constantly evolving. For the *social* dimension, the values of society also change—our understanding of development evolves, new levels of socio-economic development are achieved, and society appropriates new values and changes other values. Thus, what is socially acceptable changes over time; this is partly reflected in social movements that have appropriated emerging values such as environmental protection, more concern for greater equity in how different segments of the population are affected by current development processes, and increasing demands for citizens and their different groups to become involved in processes of planning and managing development. And, of course, the *economic* dimension is constantly changing as it reflects changing values in the marketplace, the rise of different market segments, and the changing technology of production, all of which have impacts on the economic viability of different economic activities. Inevitably, this all means that the path to sustainable development, and maintaining it, are not simple.

Because it frequently seems to be left aside in discussions about sustainable development, the environmental dimension deserves some additional comments (see Chapter 2 in particular). The natural environment has to be seen as part of human society, its communities, and its activities, particularly in rural areas. We want to conserve and protect it because of the values that it brings to all segments of society. Frequently, "conservation" and "protection" are used interchangeably, but they are not quite the same. Conservation can involve some development. If we look, for instance, at biodiversity and the efforts made to protect it, then we find that some of these efforts involve a biodiversity profile that has already been substantially influenced by human activity, particularly in territories with long histories of human settlement. From this perspective, conservation implies determining how to continue with development processes that help to maintain the valued biodiversity in a territory. This thus requires development and lies at the base of what is known as "humanized landscapes" (Domon 2009).

From the perspective of the fourth dimension of sustainable development, governance and its underlying processes, while we can investigate development from a sustainable development perspective in any national, regional, or local culture, a truly sustainable development perspective today needs to consider how the mobilization and involvement of citizens (including, of course, farmers in rural systems) are handled (see Chapter 3). Citizens can take initiatives that are central to the sustainable development of their territories and communities (see Chapters 3, 5, 6, and 7). However, the initiators of specific projects might not have as their core interests the long-term sustainability of their territories and communities. As soon as we begin to explore further the roles of different actors, we have to begin questioning the roles and motivations of these actors in rural communities.

Municipal councils and their bureaucracies, including land-use planners, have important roles to play (see Introduction), but they are not necessarily the leaders as seen from a sustainability point of view. Partly this is because, while municipal councillors are elected, they do not represent all of the legitimate interests in a community, and they might not want to either. Furthermore, land-use planners essentially can be considered as being at the service of the community. This is why, in relation to strategic development planning, it is common to talk of it *for* and *by* the community (e.g., Bryant 1999; Bryant et al. 2004; Haliburton County 2013), and it is not uncommon for strategic development planning to take on a more important role than land-use planning simply because some of the most important initiatives have little to do with land-use planning per se (e.g., improved communication between official "leaders" of a community and the population or the fact that many initiatives can be taken within land-use plans that are not really referred to in those plans, such as business incubators particularly when they are not thought of in terms of a specific building but as the whole community; this includes farm activities compatible with land-use plans but frequently not specifically identified in the plans, such as the workshops on various Mennonite farms in southwestern Ontario that have sometimes led to specific businesses being created, either formally or informally). Strategic planning for development for and by the community is essentially a continual process. An example is Haliburton County, in Ontario, long regarded (since the mid-1990s) as having a very successful approach to planning, strategically using citizen input and recognizing and cultivating citizen responsibility. This converges with much of what is said in several chapters in this book.

It is also important to recognize the territorial specificities in relation not just to climate change and adaptation and peak oil but also to all of the topics dealt with in this book. This territorialization of all aspects of rural community and activity change (e.g., farming) is central to rural sustainability because solutions to whatever challenges are faced by rural communities need to be tailored to take into account those specificities. Many of the chapters in this book, with a frequent focus on case studies, converge with this perspective.

FORCES, CONFLICTS, AND UNCERTAINTIES FACED BY RURAL SYSTEMS

When discussing the different forces of change and how they can influence rural sustainability, it is important to recognize that there are multiple forces. Although this book has an important focus on climate change and variability and peak oil, it is crucial that we recognize that many other forces or stressors

influence the many decision-making processes present in or affecting rural systems. So, though it is essential that decision makers recognize climate change
and variability and peak oil as forces that have been and likely will be affecting
rural communities and activities, the existence of multiple other forces means
that some of them are given greater priority in decision making (e.g., the globalization of markets and production processes; government decisions about
changing the "rules," such as for the horse-racing industry in Ontario since
2012; government decisions that reinforce the "attractiveness" of large urban
areas, such as the Toronto-centred region; and so on). Most of these multiple
stressors affecting rural systems have been recognized in the academic literature
(Bryant 2013; Bryant et al. 2011) and increasingly are taken into account in strategic planning for territorial development.

Some of the stressors can actually represent positive externalities for rural
systems, particularly those associated with the multiple functions of rural
communities and their land bases. Thus, maintaining the farmland resource
base can be facilitated when actors and citizens recognize the value to their
own local societies of the multiple values supported by the farmland base and
farming activities, in particular the functions of the conservation of heritage
landscapes, the maintenance of biodiversity (depending on the type of farming system), and other eco-services (see Chapter 8; see also Bryant 2011 for
farming in the Montreal region; similar functions can be associated with
many of the farming areas in the Golden Horseshoe Greenbelt; see Ontario
2004, 2005). This multifunctionality can also be important in building community resiliency and solidarity among non-farm citizens and collective
actors and farmers. Other stressors can present negative externalities, such as
water pollution from "modern" farm technology and health preoccupations
related to food quality as well as food security (see Chapter 9).

When we explore the multiple forces influencing rural systems and their
communities and activities, we cannot help but recognize the complexities
involved. As already noted, both positive and negative externalities (the latter can be related to conflicts) can arise. However, there is also considerable
uncertainty. Most decisions taken by the multiple actors in rural communities affect the future of rural systems, and often these decisions are made in a
context of uncertainty. The consequences of several of the multiple forces are
not known for certain, yet decisions must be made sooner or later (e.g., farm
diversification or not; diversification of the economic bases of rural communities and territories or not; maintenance of the level of public services or not;
improvement of transportation between rural areas and major urban areas or
not). Furthermore, we have to recognize that there is likely not a single solution simply because of the considerable heterogeneity of the rural territories
and communities even in a single province. Such uncertainty emphasizes

the need for a continuous process of reflection, which we can call a strategic reflection process and which ideally should provide the background to any planning (land use, infrastructure development such as transportation development [see Chapter 4], economic development, cultural development) in rural communities.

VULNERABILITY AND RURAL SUSTAINABILITY

When everything is going well for a rural community, things are great. But this does not mean that these communities are not vulnerable. One of the challenges is that, when things are going well, it is often not easy to get people—councillors, elected officials, citizens—to think in terms of what could happen when things start going wrong and suddenly they wake up to the fact that their community is vulnerable (see, e.g., Chapter 1).

What does vulnerability mean? Three dimensions are usually identified: exposure, sensitivity, and adaptive capacity (Delusca 2010). The vulnerability of a system (rural or urban), other than simple exposure to sources of risks, therefore relates to the sensitivity of a system (e.g., a community or territory) in the face of different sources of stress and to the difficulties of regaining a reasonable state of functioning, not necessarily of course to return to the "initial" state, which has usually contributed to the territory's or community's vulnerability (e.g., a narrow economic base when the principal activity faces a major disruption).

In Table 10.1, some of the forces that can affect the *adaptive capacity* of a rural system (communities, activities, individuals) are identified, as are others that can affect the *degree of exposure* to different sources of risk. Adaptive capacity is not simply being able to cope with a suddenly difficult situation but also being able to engage in proactive adaptation—to reduce the community's vulnerability—in anticipation of the negative consequences of particular events, such as an extreme climate event (Brklacich and Bohle 2006; Bryant, Singh, and Thomassin 2008) or the closure of a principal economic activity. Once again this raises the importance of undertaking strategic analysis and development planning, which involve placing the system under study (community, territory) into a context in which the system is compared with other communities and territories and the different stressors both present and future are anticipated as far as possible. If deemed necessary, appropriate actions can then be taken or at least planned. Inevitably, this form of reflection and development planning can have implications for land-use planning, more so in the medium to long term than in the short term. Also, developing strategies to reduce vulnerability inevitably requires co-construction of strategies and initiatives, implying the involvement of other collective actors,

FORCES AFFECTING THE VULNERABILITY OF RURAL SYSTEMS

FORCES ASSOCIATED WITH ADAPTIVE CAPACITY (COMMUNITIES, ACTIVITIES, INDIVIDUALS):	‣ the situation regarding preparedness and training; ‣ family and business finances; ‣ the dynamics of communities, including how "leaders" perceive different actors and segments of interest in the community; community support and culture; ‣ the ability of different actors and citizens to work together for a more sustainable system; and ‣ the extent to which local and regional initiatives are encouraged
FORCES THAT CAN ALTER THE DEGREE OF EXPOSURE:	‣ climate change and variability; ‣ peak oil; ‣ urbanization; ‣ technological change; ‣ globalization, market accessibility, and competition; ‣ effectiveness of information diffusion; and ‣ actions of the government (e.g., programs, policies regarding trade, policies regarding the environment and sustainable development).

Table 10.1. Forces affecting the vulnerability of rural systems (communities, territories, activities)

including citizens, frequently through the development of new citizen associations or organizations.

Adaptive capacity is a key element of building (community) resilience, and it includes the capacity to innovate. In effect, reflecting on adaptation as well as resilience should become an ongoing process helped by following what is happening to other communities in similar regions as well as other types of regions. Adaptive capacity can be explored at the level of individuals, households, and businesses as well as planning processes (land use, community economic development, strategic development planning).

COMMUNITY RESILIENCE

Community resilience is now discussed both as a framework for analysis and as a strategy to pursue and integrate into a holistic process for planning (necessarily a form of strategic planning for development) and action.

The Components of Community Resilience – Community resilience has to do with the capacity of a system (community, territory) and its subsystems (e.g., the different activities pursued) to adapt to changing circumstances, including different types of perturbations (e.g., major floods [linked to major fluctuations in climatic conditions], major catastrophes, surges in the costs of inputs for major activities [peak oil], and closures of major economic activities). Community resilience does not simply mean the capacity to adapt to these changing conditions; it also means the ability to develop a more resilient system. It can be achieved ahead of major difficulties; although it is not possible to identify all sources of difficulty ahead of time, we do know the range of forces that can lead to vulnerability. It is therefore possible first to monitor what is happening to a given community or territory and then to develop strategies to reduce negative impacts and vulnerabilities and thereby to increase a system's resilience.

In the Introduction, the topic of community resilience is broached in relation to climate change and peak oil, and much attention is given to the role of planners and municipal councils. Community resilience must also deal with the whole range of potential stressors, the vulnerability of rural communities, and issues of conflict, actual and potential, as well as uncertainty.

The main components of community resilience are social, economic, political/governance, and ecosystemic. Community resilience involves taking into account all of them and particularly how the different actors interact (the dynamics of communities, their collective actors and citizens, the nature of leadership, and the values that dominate in different communities) and look to the uncertain future. Community resilience by nature is holistic since in effect it is a tool or condition that helps a community to deal with any combination of stressors, with both positive and negative consequences.

It is important to underscore the point that, without the recognition and involvement of citizens, not much progress can be made either in resilience building or in developing adaptive capacity and reducing vulnerability. Resilience building is therefore not just a reaction to difficulties once they have occurred. In one particular action-research project (ARUC-CURA 2013), among many others, community resilience has been seen as an important strategic orientation to plan for and achieve through different initiatives. Thus, a major challenge is to ensure that building community resilience becomes part of the overall process of strategic reflection and planning for the development of a rural community, not something that arises when the community is confronted with a serious problem. This can be a major challenge, however, for many communities in which the focus is on "growth," whether it is realistic or not, and in which people do not wish to think about future difficulties. The challenge is to get everyone to appreciate as far as possible the need for co-construction of planning and action (Bryant 2010) and to bring together the different collective actors and citizens in a community to reflect together, appreciate the importance of community resilience, and encourage everyone to engage in building resilience even when things are going well. In any case, community resilience is good for the community when things are going well; community solidarity, one aspect of community resilience, can involve helping those who have become marginalized, a frequent component of community economic development (Bryant and Bruce 2009).

Thus, given the multiple external stressors currently affecting many rural communities, including the agricultural sector, not to take account of them in overall planning and management is akin to opening the community up to repeated failures. This means that community resilience must become part of the overall strategic reflection and planning approach both in relation to community development and in the planning and development of different orientations; these orientations inevitably vary among communities and territories, but they often include strategic orientations such as farming, environmental conservation, economic diversification, social integration, rural tourism development, and infrastructure development, to name but a few.

CONCLUSION

This chapter has drawn together many of the themes addressed by the other authors of this book. A view of community resilience based on local action and appreciation of the importance of community resilience has been presented, while recognizing the essential heterogeneity of rural communities in different local, cultural, and political contexts, even within a specific province, Ontario. At the same time, in the face of major stressors, the origins of which lie largely outside the specific territories or communities considered, we have emphasized the importance of taking a holistic perspective on reducing vulnerability and building adaptive capacity to ensure community resilience. Such a holistic approach is reflected in the processes of strategic reflection and planning for development *for* and *by* the community, necessitating the involvement and mobilization of the various actors and legitimate interests in the community. This holistic approach does not mean, of course, that we forget about local initiatives in the economic, social, environmental, and governance domains. As can be seen in many of the chapters of this book, local initiatives are central to many of the presentations and discussions. Launching specific projects can sometimes be the only way to spark a more holistic reflection: in communities in which people are skeptical about what can be done, showing that there are people in the community who can make things happen can kick-start a more general and all-encompassing process.

References

ARUC-CURA (Alliance Recherche Université-Communautés Communities-Universities Research Alliance). 2013. "ARUC-CURA défis des communautés côtières Coastal-Communities Challenges." Rimouski: Université du Québec à Rimouski. http://www. defisdescommunautescotieres.org/.

Brklacich, M., and H.-G. Bohle. 2006. "Assessing Human Vulnerability to Climatic Change." In *Earth System Science in the Anthropocene: Emerging Issues and Problems,* edited by T. Kraft and E. Ehliers, 51–61. Amsterdam: Springer Verlag.

Bryant, C.R. 1999. "Community-Based Strategic Planning, Mobilisation, and Action at the Edge of the Urban Field: The Case of Haliburton County." In *Progress in Research on Sustainable Rural Systems,* edited by I. Bowler, C.R. Bryant, and A. Firmino, 211–22. Série Estudos 2. Lisbon: Centro de Estudos de Geografia e Planeamento Regional, Universidade Nova de Lisboa.

———. 2010. "Co-Constructing Rural Communities in the 21st Century: Challenges for Central Governments and the Research Community in Working Effectively with Local and Regional Actors." In *The Next Rural Economies: Constructing Rural Place in Global Economies,* edited by G. Halseth, S. Markey, and D. Bruce, 142–54. Oxford: CABI Publishing.

———. 2011. "Les dynamiques des agricultures périurbaines autour de Montréal: Défis et opportunités au service de la société métropolitaine." In *Panorama des régions du Québec, édition 2011,* 13–28. Québec: Institut de la Statistique du Québec. http://www.stat.gouv. qc.ca/publications/regions/panorama.htm.

———. 2013. "Challenges for Research and Practice in the 21st Century for the Sustainability of Rural Systems." Keynote presentation at the 21st Annual Colloquium of the International Geographical Union Commission on the Sustainable Development of Rural Systems, Globalization, and New Challenges of Agricultural Systems. Nagoya, Japan.

Bryant, C.R., and D. Bruce. 2009. "Rural Economic Development: Critical Reflections on the Record and Potential Directions." In *Rural Planning and Development in Canada in the 21st Century: Challenges and Opportunities in the Context of Globalization,* edited by D.J. Douglas, 53–84.Toronto: Nelson Education.

Bryant, C.R., G. Chahine, K. Delusca, O. Daouda, M. Doyon, B. Singh, M. Brklacich, and P. Thomassin. 2011. "Adapting to Environmental and Urbanisation Stressors: Farmer and Local Actor Innovation in Urban and Periurban Areas in Canada." In *Actes du Symposium Innovation et Développement Durable dans l'Agriculture et l'Agroalimentaire— ISDA 2010,* edited by E. Coudel, H. Devautour, C.T. Soulard, and B. Hubert. Montpellier, France: CIRAD INRA SupAgro. http://www.isda2010.net/var/isda2010/storage/ original/application/e5abe6c6d5d6028e967425dd84803084.pdf.

Bryant, C.R., M. Doyon, S. Frej, D. Granjon, and C. Clément. 2004. "The Integration of Environment into Sustainable Development Practice and Discourse through Citizen Participation and the Mobilisation of Local Knowledge." In *The Regional Dimension and Contemporary Challenges to Rural Sustainability,* edited by A. de Souza Mello Bicalho and S. Hoefle, 14–25. Rio de Janeiro: Laget, Universidade Federal do Rio de Janeiro.

Bryant, C.R., N. Sanchez, K. Delusca, O. Daouda, and A. Sarr. 2013. "Metropolitan Vulnerability and Strategic Roles for Periurban Agricultural Territories in the Context of Climate Change and Variability." *Cuadernos de geografia* 22, 2: 58–68.

Bryant, C.R., B. Singh, and P. Thomassin. 2008. *Evaluation of Agricultural Adaptation Processes and Adaptive Capacity to Climate Change and Variability: The Co-Construction of New Adaptation Planning Tools with Stakeholders and Farming Communities in the Saguenay-Lac-Saint-Jean and Montérégie Regions of Québec.* Research report for Project A1332 submitted to Natural Resources Canada CCIAP, Ottawa.

Delusca, K. 2010. "La vulnérabilité des exploitations agricoles au Québec face au changement et à la variabilité climatiques." PhD diss., l'Université de Montréal.

Domon, G., ed. 2009. *Le paysage humanisé au Québec.* Montréal: Les Presses de l'Université de Montréal.

Haliburton County. 2013. "Strategic Plan of Haliburton County." http://www.haliburton-county.ca/services/planning-and-gis/strategic-plan/.

Ontario. Ministry of Municipal Affairs and Housing. 2004. *Greenbelt Task Force Discussion Paper: Toward a Golden Horseshoe Greenbelt: A Framework for Consultation.* Toronto: Ontario Ministry of Municipal Affairs and Housing. http://www.mah.gov.on.ca/Page1399.aspx.

———. 2005. *The Greenbelt Plan (2005).* Toronto: Ontario Ministry of Municipal Affairs and Housing. http://www.mah.gov.on.ca/.

CONTRIBUTORS

Jennifer Ball holds a PhD in rural studies, with a focus on sustainable rural communities. She has conducted research on issues of intercultural communication in professional planning, conflict management, storytelling, rural land-use planning, and the role of women in community-based peacebuilding. She is co-author of *Doing Democracy with Circles: Engaging Communities in Public Planning.* Jennifer is an accredited land-use and community development planner. Currently, Jennifer is an adjunct professor at the University of Guelph and a private consultant.

Christopher Bryant was a full professor (until August 2014) and is currently adjunct professor and director of the Laboratory on Sustainable Development and Territorial Dynamics, Geography, University of Montreal, and adjunct professor, School of Environmental Design and Rural Development, University of Guelph.

Wayne Caldwell is professor in rural planning and Director of the School of Environmental Design and Rural Development at the University of Guelph, Guelph Ontario, Canada. He is a passionate advocate for the future of rural communities. His research and practice includes the use of community-based approaches to plan for the social, environmental and economic health of rural communities. He has served as chair or president of a number of local, provincial and national organizations. This is his eighth book.

John Devlin is graduate coordinator of the Rural Planning Program in the School of Environmental Design and Rural Development, University of Guelph. He teaches planning and development theory, rural development administration, project development, and environment and development. His primary areas of academic interest include the role of the state in development; environmental assessment and public participation; good governance; and agricultural and environmental policy. He is currently involved in several funded research projects examining collaborative regional economic development initiatives, local food systems, community-based water monitoring, and maple syrup production. He is team lead for the Natural Resources Development Theme of the Rural Policy Learning Commons Partnership Project. He has conducted research or provided consulting services in Honduras, Bolivia, the Philippines, Vietnam, Egypt, Mozambique, South Africa, Zambia, Zimbabwe and Ukraine. He is co-chair of the public participation section of the

International Association for Impact Assessment and a member of the board of directors of the Organic Council of Ontario.

Erica Ferguson has worked in the non-profit sector as an executive director of a grassroots organization, and as a resource and program planner in an international charity that supports food-system change. In addition, for over ten years Erica has worked with Eko Nomos Program Development Consultants, as she has a passion for facilitating group processes to achieve and measure shared outcomes. With a MSc in Rural Planning, Erica brings specific planning skills to bear on projects that address the nexus of social justice, ecological responsibility, and economic realities as these relate to community health and resilience.

Paul Kraehling is currently a rural studies PhD student at the University of Guelph's School of Environmental Design and Rural Development. His main research area is the use of "green infrastructure" as a foundational planning tool for healthy rural communities. Paul has been exploring notions of community sustainability for most of his life, with over thirty years of professional planning experience working in various Ontario municipalities. He is a registered professional planner with the Canadian Institute of Planners, and is a member of the Ontario Professional Planners Institute.

Emanuèle Lapierre-Fortin holds a master's degree in rural planning and development from the University of Guelph, as well as an honours bachelor of arts in international development and economics from the University of Toronto. She has been working in sustainable community economic development for over six years, including three years as a worker member for Niska, a local and regional development consulting cooperative based in Sherbrooke, Quebec. She has been active on the Transition Guelph Steering Committee and provided four Trainings for Transition in Quebec. Her passions include participatory community planning processes, applied research, and social enterprise development.

Sally Ludwig is a founding member of Transition Guelph (www.transitionguelph.org) and lead certified transition trainer for central Ontario, with particular interests in effective collaboration, communication, and the inner dimensions of personal and community resilience. She is a member facilitator of the Work that Reconnects and works as a relational therapist, instructor, and clinical supervisor in couple and family therapy, group facilitator, consultant, and trainer.

Eric Marr is a PhD student in environmental economics and environmental management at the University of York in the United Kingdom. He is a graduate of the master of science program in rural planning and development at the

University of Guelph, where he conducted his thesis research on unmet transportation needs and public transportation opportunities in rural Ontario. Prior to pursuing his PhD, he was a policy advisor in rural development policy with the Ontario Ministry of Agriculture, Food and Rural Affairs.

Ralph C. Martin grew up on a beef and hog farm in Wallenstein, Ontario. He learned what is essential about agriculture from his grandfather, before he died when Ralph was seven. After 4-H, his formal education includes a BA and an MSc in biology from Carleton University and a PhD in plant science from McGill University. His love of teaching grew unexpectedly when he began teaching at the Nova Scotia Agricultural College, in 1990, and realized how students teach him too. In 2001, he founded the Organic Agriculture Centre of Canada to coordinate university research and education pertaining to organic systems across Canada. In 2011, he was appointed as professor and the Loblaw Chair in Sustainable Food Production at the Ontario Agricultural College, University of Guelph.

Tony McQuail is a farmer, environmentalist, and politician. He is a graduate of the University of Waterloo in environmental studies. He has been active in farm organizations at the local and provincial level. He is a founding member of the Ecological Farmers Association of Ontario and has been farming organically since the mid 1970s. He served as executive assistant to the Ontario Minister of Agriculture, Food and Rural Affairs in the early 1990s. He and his wife, Fran, own Meeting Place Organic Farm and operate it with assorted interns, apprentices, and Belgian Work Horses. Tony is a Holistic Management Certified Educator, helping farmers learn how to manage their land for healthy people, land, and profits.

Susanna Reid is a planner at the Huron County Planning and Development Department in Goderich, Ontario. She has a master's degree in rural planning from the University of Guelph. In both her professional and personal life she practises building community resiliency.